The R.A.M.S. Library of Alchemy

Volume 50

Transcendent Magic

by

Éliphas Lévi

R.A.M.S. Publishing Company

Transcendent Magic

Transcendent Magic

by

Éliphas Lévi

Produced by

Restorers of Alchemical Manuscripts Society

R.A.M.S. Publishing Company

R.A.M.S. Publishing Company
7309 East 102nd Street
Kansas City Missouri 64134

The R.A.M.S. Library of Alchemy, Volume 50:
Transcendent Magic
Copyright © 2018 Althea Productions LLC

All rights reserved. No part of this publication may be reproduced or transmitted in any form or by any means, electronic or mechanical, including but not limited to any information storage and retrieval system, without written permission from
Althea Productions LLC.
Reviewers may quote brief passages.

First Edition 2018

ISBN-13: 978-1725732490
ISBN-10: 1725732491

Image Processing by Philip N. Wheeler

Printed in the United States of America

Table of Contents

Disclaimer 9

Introduction 10

Biographical Preface 12

Introduction 26

THE CANDIDATE 48

THE PILLARS OF THE TEMPLE 58

THE TRIANGLE OF SOLOMON 65

THE TETRAGRAM 72

THE PENTAGRAM 81

MAGICAL EQUILIBRIUM 89

THE FIERY SWORD 97

REALIZATION 102

INITIATION 110

THE KABBALAH 114

THE MAGIC CHAIN 125

THE GREAT WORK 135

NECROMANCY 141

TRANSMUTATIONS 151

BLACK MAGIC 158

BEWITCHMENTS 161

ASTROLOGY 171

CHARMS AND PHILTRES 179

THE STONE OF THE PHILOSOPHERS – ELAGABALUS 188

THE UNIVERSAL MEDICINE 193

DIVINATION 197

SUMMARY AND GENERAL KEY OF THE FOUR SECRET SCIENCES
....203

Part II: The Ritual of Transcendent Magic 208

Introduction to Part II 210

Chapter I: Preparations 227

Chapter II: Magical Equilibrium 237

Chapter III: The Triangle of Pantacles 245

Chapter IV: The Conjuration of the Four 254

Chapter V: The Blazing Pentagram 266

Chapter VI: The Medium and the Mediator 271

Chapter VII: The Septenary of Talismans 277

Chapter VIII: A Warning to the Imprudent 293

Chapter IX: The Ceremonial of Initiates 297

Chapter X: The Key of Occultism 303

Chapter XI: The Triple Chain 308

Chapter XII: The Great Work 313

Chapter XIII: Necromancy 320

Chapter XIV: Transmutations 333

Chapter XV: The Sabbath of the Sorcerers 340

Chapter XVI: Witchcraft and Spells 360

Chapter XVII: The Writing of the Stars 368

Chapter XVIII: Philtres and Magnetism 383

Chapter XIX: The Mastery of the Sun 393

Chapter XX: The Thaumaturge..398

Chapter XXI: The Science of Prophets 406

Chapter XXII: The Book of Hermes..416

THE NUCTEMERON OF APOLLONIUS OF TYANA 454

A Word from the Publisher 472

The R.A.M.S. Library of Alchemy 473

Transcendent Magic

Dedicated to Hans W. Nintzel,
American Alchemist
and
Founder of the
Restorers of Alchemical Manuscripts Society
(R.A.M.S.)

Disclaimer

Liability: The publisher does not warrant or assume any legal liability or responsibility for the accuracy, completeness, or usefulness of any information, apparatus, product, or process disclosed. The publisher makes no representation as to the accuracy or completeness of the contents of this book and specifically disclaims any implied warranty of merchantability or fitness for a particular purpose. No warranty may be created or extended by written sales materials or sales representatives. You should obtain professional consultation where appropriate. The publisher shall not be liable for any loss of profit or other commercial or personal damages, including but not limited to special, incidental, consequential, or other damages.

This book is sold for informational purposes only. Neither the publisher nor the editor shall be held accountable for the use or misuse of the information in this book.

Introduction
Philip N. Wheeler

Lévi writes, "Magic is the traditional science of the secrets of nature which has been transmitted to us from the magi." Since this coincides with the general goals of Alchemy, and due to the many references to alchemical subjects contained herein, I have included it as part of the R.A.M.S. Library of Alchemy.

"To be ever rich, to be always young and to die never: such, from all time, has been the dream of alchemists," *Ritual*, Chapter XII: The Great Work. Is this not the dream of most people, and not only alchemists?

Dogma et Rituel de la Haute Magie was written by Éliphas Lévi (Éliphas Lévi Zahed, born Alphonse Louis Constant (February 8, 1810 – May 31, 1875)). Lévi was a French Occult author and ceremonial magician. The text was translated from French into English by Arthur Edward Waite (1857-1942). Although the book's title was "Transcendental Magic," within the text the work is referred to as "Transcendent Magic."

This text was meticulously gathered from several sources, with the copy in the University of Toronto Library being the primary source. There were some damaged pages in that copy, and while comparing several editions to locate the missing pages I found that some of the text had been changed or even omitted between various printings. In particular, certain of the paragraphs in sections on so-called Black Magic were missing in some copies. I here present the complete edition based on all source materials available to me.

The writings of Éliphas Lévi are an important part of the teachings of The Golden Dawn. Israel Regardie wrote, "Prominent among the Adepti of our Order and of public renown were Éliphas Lévi the greatest of modern French magi."[1] His discussions of the Tarot are most interesting, and coincide with those presented by Israel Regardie in his Golden Dawn series.

This edition includes Lévi's translation of the "Nuctemeron of Apollonius of Tyana," a short but very interesting work in itself.

---

[1] *The Golden Dawn* Volume 1 by Israel Regardie, 1937.

Biographical Preface
Arthur Edward Waite

Éliphas Lévi Zahed is a pseudonym which was adopted in his occult writings by Alphonse Louis Constant, and it is said to be the Hebrew equivalent of that name. The author of *Dogme et Rituel de la Haute Magie* was born in humble circumstances about the year 1810, being the son of a shoemaker. Giving evidence of unusual intelligence at an early age, the priest of his parish conceived a kindly interest for the obscure boy, and got him on the foundation of Saint Sulpice, where he was educated without charge, and with a view to the priesthood. He seems to have passed through the course of study at that seminary in a way which did not disappoint the expectations raised concerning him. In addition to Greek and Latin, he is believed to have acquired considerable knowledge of Hebrew, though it would be an error to suppose that any of his published works exhibit special linguistic attainments. He entered on his clerical novitiate, took minor orders, and in due course became a deacon, being thus bound by a vow of perpetual celibacy. Shortly after this step, he was suddenly expelled from Saint Sulpice for holding opinions contrary to the teaching of the Roman Catholic Church. The existing accounts of this expulsion are hazy, and incorporate unlikely elements, as, for example, he was sent by his ecclesiastical superiors to take duty in country places, where he preached with great eloquence what, however, was doctrinally unsound; but I believe that there is no precedent for the preaching of deacons in the Latin Church. Pending the appearance of the biography which has been promised for some years in France, we have few available materials for a life of the "Abbé" Constant. In any case, he was cast back upon the world, with the limitations of priestly engagements, while the priestly career was closed to him – and what he did, or how he contrived to support himself, is unknown. By the year 1839 he had made some literary friendships, including that of

Alphonse Esquiros, the forgotten author of a fantastic romance, entitled "The Magician"[2]; and Esquiros introduced him to Ganneau, a distracted prophet of the period, who had adopted the dress of a woman, abode in a garret, and there preached a species of political illuminism, which was apparently concerned with the restoration of *la vraie légitimité*. He was, in fact, a second incarnation of Louis XVII. – "come back to earth for the fulfillment of a work of regeneration."[3] Constant and Esquiros, who had visited him for the purpose of scoffing, were carried away by his eloquence, and became his disciples. Some element of socialism must have combined with the illuminism of the visionary, and this appears to have borne fruit in the brain of Constant, taking shape ultimately in a book or pamphlet, entitled "The Gospel of Liberty," to which a transient importance was attached, foolishly enough, by the imprisonment of the author for a term of six months. There is some reason to suppose that Esquiros had a hand in the production, and also in the penalty. His incarnation over, Constant came forth undaunted, still cleaving to his prophet, and undertook a kind of apostolic mission into the provinces, addressing the country people, and suffering, as he himself tells us, persecution from the ill-disposed.[4] But the prophet ceased to prophesy,

---

[2] M. Papus, a contemporary French occultist, in an extended study of the "Doctrine of Éliphas Lévi," asks scornfully" "Who now remembers anything of Paul Augnez or Esquiros, journalists pretending to initiation, and posing as professors of the occult sciences in the salons they frequented?" No doubt they are forgotten, but Éliphas Lévi states, in the *Histoire de la Magie*, that, by the of his romance of "The Magician," Esquiros founded a new school of fantastic magic, and gives sufficient account of his work to show that it was in parts excessively curious.

[3] A woman who was associated with his mission, was, in like manner, supposed to have been Marie Antoinette. – See *Histoire de la Magie*, l. 7., c. 5.

[4] A vicious story, which has received recently some publicity in Paris, charges Constant with spreading a report of his death soon after his release from prison, assuming another name, imposing upon the Bishop of Eveux, and obtaining a license to preach and administer the sacraments in that diocese, although he was not a priest. He is represented as drawing large congregations to the cathedral by his preaching, but at length the judge who had sentenced him unmasked the imposter, and the sacrilegious farce thus terminated dramatically.

presumably for want of an audience, and *la vraie légitimité* was not restored, so the disciple returned to Paris, where, in spite of the pledge of his diaconate, he effected a runaway match with Mdlle. Noémy, a beautiful girl of sixteen. This lady bore him two children, who died in tender years, and subsequently she deserted him. Her husband is said to have tried all expedients to procure her return,[5] but in vain, and she even further asserted her position by obtaining a legal annulment of her marriage, on the ground that the contracting parties were a minor and a person bound to celibacy by an irrevocable vow. The lady, it may be added, had other domestic adventures, ending in a second marriage about the year 1872. Madame Constant was not only very beautiful, but exceedingly talented, and after her separation she became famous as a sculptor, exhibiting at the Salon and elsewhere under the name of Claude Vingmy. It is not impossible that she may be still alive; in the sense of her artistic genius, at least, she is something more than a memory.

At what date Alphonse Louis Constant applied himself to the study of the occult sciences is uncertain, like most other epochs of his life. The statement that in the year 1825 he entered on a fateful path, which led him through suffering to knowledge, must not be understood in the sense that his initiation took place at that period, which was indeed early in boyhood. It obviously refers to his enrollment among the scholars of Saint Sulpice, which, in a sense, led to suffering, and perhaps ultimately to science, as it certainly obtained him education. The episode of the New Alliance – so Gannean termed his system – connects with transcendentalism, at least on the side of hallucination, and may have furnished the required impulse to the mind of the disciple; but in 1846 and 1847, certain pamphlets issued by Constant under the auspices of the *Libraire Sociétaire* and the *Libraire*

---

[5] Including Black Magic and pacts with Lucifer, according to the silly calumnies of his enemies.

Phalanstérienne shew that his inclinations were still towards Socialism, tinctured by religious aspirations. The period which intervened between his wife's desertion[6] and the publication of the *Dogme de la Haute Magie*, in 1855, was that, probably, which he devoted less or more to occult study. In the interim he issued a large "Dictionary of Christian Literature," which is still extant in the encyclopædic series of Abbé Migne; this work betrays no leaning towards occult science, and, indeed, no acquaintance therewith. What it does exhibit unmistakably is the intellectual insincerity of the author, for which he assumes therein the mask of perfect orthodoxy, and that accent in matters of religion which is characteristic of the voice of Rome. The *Dogme de la Haute Magie* was succeeded in 1856 by its companion volume *Rituel*, both of which are here translated for the first time into English. It was followed in rapid succession by the *Histoire de la Magie*, 1860; *La Clef des Grands Mystères*, 1861; a second edition of the *Dogme et Rituel*, to which a long and irrelevant introduction was unfortunately prefixed, 1862; *Fables et Symboles*, 1864; *Le Sorcier de Meudon*, a beautiful pastoral idyll, impressed with the *cachet cabalistique*; and *La Science des Espirits*, 1865. The last two works incorporate the substance of the pamphlets published in 1846 and 1847.

The precarious existence of Constant's younger days was in one sense but faintly improved in his age. His books did not command a large circulation, but they secured him admirers and pupils, from whom he received remuneration in return for personal or written courses of instruction. He was commonly to be found *chez lui* in a species of magical vestment, which may be pardoned in a French magus, and his only available portrait – prefixed to this volume – represents him in

---

[6] I must not be understood as definitely attaching blame to Madame Constant for the course she adopted. Her husband was approaching middle life when he withdrew her – still a child – from her legal protectors, and the runaway marriage which began by foreswearing was, under the circumstances, little better than a seduction thinly legalized, and it was afterwards not improperly dissolved.

that guise. He outlived the Franco-German war, and as he had exchanged Socialism for a sort of transcendentalized Imperialism, his political faith must have been as much tried by the events which followed the siege of Paris as was his patriotic enthusiasm by the reverses which culminated in Sédan. His contradictory life closed in 1875 amidst the last offices of the church which had almost expelled him from her bosom. He left many manuscripts behind him, which are still in course of publication, and innumerable letters to his pupils – Baron Spedalieri alone possesses nine volumes – have been happily preserved in most cases, and are in some respects more valuable than the formal treatises.

No modern expositor of occult science can bear any comparison with Éliphas Lévi, and among ancient expositors, though many stand higher in authority, all yield to him in living interest, for he is actually the spirit of modern thought forcing an answer for the times from the old oracles. Hence there are greater names, but there is no influence so great – no fascination in occult literature exceeds that of the French magus. The others are surrendered to specialists and the typical serious students to whom all dull and unreadable masterpieces are dedicated, directly or not; but he is read and appreciated, much as we read and appreciate new and delightful verse which, through some conceit of the poet, is put into the vesture of Chaucer. Indeed, the writings of Éliphas Lévi stand, as regards the grand old line of initiation, in relatively the same position as the "Earthly Paradise" or Mr. William Morris stands to the "Canterbury Tales." There is the recurrence to the old conceptions, and there is the assumption of the old drapery, but there is in each case the new spirit. The "incommunicable axiom" and the "great Arcanum," Azoth, Inri, Tetragrammaton, which are the vestures of the occult philosopher, are like the "cloth of Bruges and hogsheads of Guienne, Florence gold cloth, and Ypres napery" of the poet. In both cases it is the year 1850 *et seq.*, in a mask of high fantasy. Moreover, "the idle singer of an empty day" is paralleled fairly enough by "the poor and obscure scholar who has recovered the lever of Archimedes." The comparison is intentionally grotesque, but it obtains notwithstanding, and even admits of development, for as Mr. Morris in a sense voided the *reason d'étre* of his poetry, and, in express contradiction to his own mournful question, has endeavored to "set the crooked straight" by betaking himself to Socialism, so Éliphas Lévi surrendered the rod of miracles and voided his Doctrine of Magic by devising a one-sided and insincere concordat with orthodox religion, and expiring in the arms of "my venerable masters in theology," the descendants, and decadent at that, of the "imbecile theologians of the middle ages." But the one is, as the other was, a man of sufficient ability to make a paradoxical defense of a position which remains untenable.

Students of Éliphas Lévi will be acquainted with the qualifications and stealthy retractions by which the uncompromising position of initiated superiority in the "Doctrine and Ritual," had its real significance read out of it by the later works of the magus. I have dealt with this point exhaustively in another place,[7] and there is no call to pass over the same ground a second time. I propose rather to indicate as briefly as possible some new considerations which will help us to understand why there were grave discrepancies between the "Doctrine and Ritual of Transcendental Magic" and the volumes which followed these. In the first place, the earlier books were written more expressly from the standpoint of initiation, and in the language thereof; they obviously contain much which it would be mere folly to construe after a literal fashion, and what Éliphas Lévi wrote at a later period is not so much discrepant with his earlier instruction – though it is this also – as the qualifications placed by a modern transcendentalist on the technical

---

[7] See the Critical Essay prefixed to *The Mysteries of Magic: a Digest of the Writings of Éliphas Lévi*. London: George Redway. 1886.

exaggerations of the secret sciences. For the proof we need travel no further than the introduction to "The Doctrine of Magic," and to the Hebrew manuscript cited therein, as to the powers and privileges of the magus. Here the literal interpretation would be insanity; these claims conceal a secret meaning, and are trickery in their verbal sense. They are what Éliphas Lévi himself terms "hyperbolic," adding: "If the sage do not materially and actually perform these things, he accomplishes others which are much greater and more admirable." But this consideration is not in itself sufficient to take account of the issues that are involved; it will not explain, for example, why Éliphas Lévi, who consistently teaches in the "Doctrine and Ritual" that the dogmas of so-called revealed religion are nurse-tales for children, should subsequently have insisted on their acceptation in the sense of the orthodox Church by the grown men of science, and it becomes necessary here to touch upon a matter which, by its nature, and obviously, does not admit of complete elucidation.

The precise period of study which produced the "Doctrine and Ritual of Transcendent Magic" as its first literary result is not indicated with any certainty, as we have seen in the life of the author, nor do I regard Éliphas Lévi as constitutionally capable of profound or extensive book study. Intensely suggestive, he is at the same time without much evidence of depth; splendid in generalization, he is without accuracy in detail, and it would be difficult to cite a worse guide over mere matters of fact. His "History of Magic" is a case in point; as a philosophical survey it is admirable, and there is nothing in occult literature to approach it for literary excellence, but it swarms with historical inaccuracies; it is in all respects an accomplished and in no way an erudite performance, nor do I think that the writer much concerned himself with any real reading of the authorities whom he cites. The French verb *parcourir* represents his method of study, and not the verb *approfondir*. Let us take one typical case. There is no occult writer whom he cites with more satisfaction, and towards whom he exhibits more reverence, than William Postel, and of all Postel's books there is none which he mentions so often as the *Clavis Absconditorum à Constitutione Mundi*; yet he had read this minute treatise so carelessly that he missed a vital point concerning it, and apparently died unaware that the symbolic key prefixed to it was the work of the editor and not the work of Postel. It does not therefore seem unreasonable to affirm that had Lévi been left to himself, he would not have got far in occult science, because his Gallic vivacity would have been blunted too quickly by the horrors of mere research; but he did somehow fall within a circle of initiation which curtailed the necessity for such research, and put him in the right path, making visits to the Bibliothèque Nationale and the Arsenal of only subsidiary importance. This, therefore, constitutes the importance of the "Doctrine and Ritual"; disguised indubitably, it is still the voice of initiation; of what school does not matter, for in this connection nothing can be spoken plainly, and I can ask only the lenience of deferred judgement from my readers for my honorable assurance that I am not speaking idly. The grades of that initiation had been only partly ascended by Éliphas Lévi when he published the *Doctrine and Ritual*, and its publication closed the path of his progress: as he was expelled by Saint Sulpice for the exercise of private judgment in matters of doctrinal belief, so he was expelled by his occult chiefs for the undue exercise of personal discretion in the matter of the revelation of the mysteries. Now, these facts explain in the first place the importance, as I have said, of the *Doctrine and Ritual*, because it represents a knowledge which cannot be derived from books; they explain, secondly, the shortcomings of that work, because it is not the result of a full knowledge; why, thirdly, the later writings contain no evidences of further knowledge; and, lastly, I think that they materially assist us to understand why there are retractions, qualifications, and subterfuges in the said later works. Having gone too far, he naturally attempted to go back, and just as he strove to patch up a species of *modus vivendi* with the church of his childhood, so he endeavored, by throwing dust in the eyes of his readers, to make peace with that initiation, the first law of which he had indubitably violated. In both cases, and quite naturally, he failed.

It remains for me to state what I feel personally to be the chief limitation of Lévi, namely, that he was a transcendentalist but not a mystic, and, indeed, he was scarcely a transcendentalist in the accepted sense, for he was fundamentally a materialist – a materialist, moreover, who at times approached perilously towards atheism, as when he states that God is a hypothesis which is "very probably necessary"; he was, moreover, a disbeliever in any real communication with the world of spirits. He defines mysticism as the shadow and the buffer of intellectual light, and loses no opportunity to enlarge upon its false illuminism, its excesses, and fatuities. There is, therefore, no way from man to God in his system, while the sole avenues of influx from God to man are sacramentally, and in virtue merely of a tolerable hypothesis. Thus man must remain in simple intellectualism if he would rest in reason; the sphere of material experience is that of his knowledge; and as to all beyond it, there are only the presumptions of analogy. I submit that this is not the doctrine of occult science, nor the *summum bonum* of the greater initiation; that transcendental pneumatology is more by its own hypothesis than an alphabetical system argued kabbalistically; and that more than mere memories can on the same assumption be evoked in the astral light. The hierarchic order of the physical world has its complement in the invisible hierarchy, which analogy leads us to discern, being at the same time a process of our perception rather than a rigid law governing the modes of manifestation in all things seen and unseen; initiation takes us to the bottom step of the ladder of the invisible hierarchy and instructs us in the principles of ascent, but the ascent rests personally with ourselves; the voices of some who have preceded can be heard above us, but they are of those who are still upon the way, and they die as they rise into the silence, towards which we also must ascend alone, where initiation can no longer help us, unto that bourne from whence no traveler returns, and the influxes are sacramental only to those who are below.

An annotated translation exceeded the scope of the present undertaking, but there is much in the text which follows that offers scope for detailed criticism, and there are points also where further elucidation would be useful. One of the most obvious defects, the result of mere carelessness or undue haste in writing, is the promise to explain or to prove given points later on, which are forgotten subsequently by the author. Instances will be found concerning the method of determining the appearance of unborn children by means of the pentagram; concerning the rules for the recognition of sex in the astral body; concerning the notary art; concerning the magical side of the Exercises of St. Ignatius; concerning the alleged sorcery of Grandier and Girard; concerning Schrœpffer's secrets and formulas for evocation; concerning the occult iconography of Gaffarel. In some cases the promised elucidations appear in other places than those indicated, but they are mostly wanting altogether. There are other perplexities with which the reader must dead according to his judgement. The explanation of the quadrature of the circle is a childish folly; the illustration of perpetual motion involves a mechanical absurdity; the doctrine of the perpetuation of the same physiognomies from generation to generation is not less absurd in heredity; the cause assigned to cholera and other ravaging epidemics, more especially the reference to bacteria, seems equally outrageous in physics. There is one other matter to which attention should be directed; the Hebrew quotations in the original – and the observation applies generally to all the works of Lévi – swarm with typographical and other errors, some of which it is impossible to correct, as, for example, the passage cited from Rabbi Abraham. So also the Greek conjuration is simply untranslatable as it stands, and the version given is not only highly conjectural, but omits an entire passage owing to insuperable difficulties. Lastly, after careful consideration. I have judged it the wiser course to leave out the preliminary essay which was prefixed to the second edition of the "Doctrine and Ritual;" its prophetic utterances upon the mission of Napoleon III have been stultified by subsequent events; it is devoid of any connection with the work which it precedes, and, representing as it does the later views of Lévi, it would be a source of confusion to the reader. The present translation represents, therefore, the first edition of the *Dogme et Rituel de la Haute Magie,* omitting nothing but a few unimportant citations from old French grimoires in an unnecessary appendix at the end. The portrait of Lévi is from a *carte-de-visite* in the possession of Mr. Edward Maitland, and was issued with his *Life of Anna Kingsford*, a few months ago.

A. E. Waite, London, September 1896.

Éliphas Lévi

Transcendent Magic

Introduction

Behind the veil of all the hieratic and mystical allegories of ancient doctrines, behind the shadows and the strange ordeals of all initiations, under the seal of all sacred writings, in the ruins of Nineveh or Thebes, on the crumbling stones of the old temples, and on the blackened visage of the Assyrian or Egyptian sphinx, in the monstrous or marvelous paintings which interpret to the faithful of India the inspired pages of the Vedas, in the strange emblems of our old books of alchemy, in the ceremonies at reception practiced by all secret societies, traces are found of a doctrine which is everywhere the same, and everywhere carefully concealed. Occult philosophy seems to have been the nurse or god-mother of all intellectual forces, the key of all divine obscurities, and the absolute queen of society in those ages when it was reserved exclusively for the education of priests and of kings. It reigned in Persia with the magi, who at length perished, as perish all masters of the world, because they abused their power; it endowed India with the most wonderful traditions, and with an incredible wealth of poesy, grace and terror in its emblems; it civilized Greece to the music of the lyre of Orpheus; it concealed the principles of all the sciences and of all human intellectual progress in the bold calculations of Pythagoras; fable abounded in its miracles, and history, attempting to appreciate this unknown power, became confused with fable; it shook or strengthened empires by its oracles, caused tyrants to tremble on their thrones, and governed all minds, either by curiosity or by fear. For this science, said the crowd, there is nothing impossible, it commands the elements, knows the language of the stars, and directs the planetary courses; when it speaks, the moon falls blood-red from heaven; the dead rise in their graves and articulate ominous words as the night wind blows through their skulls. Mistress of love or of hate, occult science can dispense paradise or hell at its pleasure to human hearts; it disposes of all forms and distributes beauty or ugliness; with the rod of Circe it alternately changes men into brutes and animals into men; it even disposes of life and death, can confer wealth on its adepts by the transmutation of metals and immortality by its quintessence or elixir compounded of gold and light. Such was magic from Zoroaster to Manes, from Orpheus to Apollonius of Tyana, when positive Christianity, at length victorious over the brilliant dreams and titanic aspirations of the Alexandrian school, dared to launch its anathemas publicly against this philosophy, and thus forced it to become more occult and mysterious than ever. Moreover, strange and alarming rumors began to circulate concerning initiates or adepts; these men were everywhere surrounded by an ominous influence; they killed or drove mad those who allowed themselves to be carried away by their honeyed eloquence or by the fame of their learning. The women whom they loved became Stryges, their children vanished at their nocturnal meetings, and men whispered shudderingly and in secret of bloody orgies and abominable banquets. Bones had been found in the crypts of ancient temples, shrieks had been heard in the night, harvests withered and herds sickened when the magician passed by. Diseases which defied medical skill at times appeared in the world, and always, it was said, beneath the envenomed glance of the adepts. At length a universal cry of execration went up against magic, the mere name became a crime and the common hatred was formulated in this sentence: "Magicians to the flames!" as it was shouted some centuries earlier: "To the lions with the Christians!" Now the multitude never conspires except against real powers; it possesses not the knowledge of what is true, but it has the instinct of what is strong. It remained for the eighteenth century to deride both Christians and magic, while infatuated with the disquisitions of Rousseau and the illusions of Cagliostro.

Science, notwithstanding, is at the basis of magic, as at the foundation of Christianity there is love, and in the Gospel symbols we see the Word incarnate adored in his cradle by three magi, led thither by a star (the triad and the sign of the microcosm) and receiving their gifts of gold, frankincense and myrrh, a second mysterious triplicity, under which emblem the highest secrets of the Kabbalah are allegorically contained.

Christianity owes, therefore, no hatred to magic, but human ignorance has ever stood in fear of the unknown. The science was driven into hiding to escape the impassioned assaults of blind love: it clothed itself with new hieroglyphics, dissimulated its labors, denied its hopes. Then it was that the jargon of alchemy was created, a permanent deception for the vulgar, a living language only for the true disciple of Hermes.

Extraordinary fact! Among the sacred records of the Christians there are two works which the infallible Church makes no claim to understand and has never attempted to explain: these are the prophecy of Ezekiel and the *Apocalypse*, two Kabbalistic keys assuredly reserved in heaven for the commentaries of magician Kings, books sealed as with seven seals for faithful believers, yet perfectly plain to an initiated infidel of the occult sciences. There is also another book, but, although it is popular in a sense and may be found everywhere, this is of all most occult and unknown, because it has the key of all others; it is in public evidence without being known to the public; no one dreams of seeking it where it actually is, and elsewhere it is lost labor to look for it. This book, possibly anterior to that of Enoch, has never been translated, but is still preserved unmutilated in primeval characters, on detached leaves, like the tablets of the ancients. A distinguished scholar has revealed, though no one has observed it, not indeed its secret, but its antiquity and singular preservation; another scholar, but of a mind more fantastic than judicious, passed thirty years in the study of this book, and has merely suspected its whole importance. It is, in fact, a monumental and extraordinary work, strong and simple as the architecture of the pyramids, and consequently enduring like those – a book which is the sum of all the sciences, which can resolve all problems by its infinite combinations, which speaks by evoking thought, is the inspirer and regulator of all possible conceptions, the masterpiece perhaps of the human mind, assuredly one of the finest things bequeathed to us by antiquity, a universal key, the name of which has been explained and comprehended only by the learned William Postel, an unique text, whereof the initial characters alone exalted the devout spirit of Saint

Martin into ecstasy, and might have restored reason to the sublime and unfortunate Swedenborg. We shall speak of this book later on, and its mathematical and precise explanation will be the complement and crown of our conscientious undertaking. The original alliance of Christianity and the science of the magi, once it is thoroughly demonstrated, will be a discovery of no second-rate importance, and we question not that the serious study of magic and the Kabbalah will lead earnest minds to the reconciliation of science and dogma, of reason and faith, heretofore regarded as impossible.

We have said that the Church, whose special office is the custody of the Keys, does not pretend to possess those of the *Apocalypse* or of Ezekiel. In the opinions of Christians, the scientific and magical clavicles of Solomon are lost; yet, at the same time, it is certain that, in the domain of the intelligence ruled by the Word, nothing which has been written can perish; things which men cease to understand simply cease to exist for them, at least in the order of the Word, and they enter then into the domain of enigma and mystery. Furthermore, the antipathy, and even open war, of the official church against all that belongs to the realm of magic, which is a kind of personal and emancipated priesthood, is allied with necessary and even with inherent causes in the social and hierarchic constitution of Christian sacerdotalism. The Church ignores magic – for she must either ignore it or perish, as we shall prove later on; yet she does not the less recognize that her mysterious founder was saluted in his cradle by the three magi – that is to say, by the hieratic ambassadors of the three parts of the known world and the three analogical worlds of occult philosophy. In the school of Alexandria, magic and Christianity almost joined hands under the auspices of Ammonius Saccas and of Plato; the doctrine of Hermes is found almost in its entirety in the writings attributed to Denis the Areopagite; and Synesius sketched the plan of a treatise on dreams, which was later on to be annotated by Cardan, and composed hymns which might have served for the liturgy of the Church of Swedenborg, could a church of the illuminated possess a liturgy. With this period of fiery abstractions and impassioned warfare of words there must also be connected the philosophic reign of Julian, called the Apostate because in his youth he made an unwilling profession of Christianity. Everyone is aware that Julian was sufficiently wrongheaded to be an unseasonable hero of Plutarch, and was, if one may say so, the Don Quixote of Roman Chivalry; nut what most people do not know is that Julian was one of the illuminated and an initiate of the first order; that he believed in the unity of God and in the universal doctrine of the Trinity; that, in a word, he regretted nothing of the old world but its magnificent symbols and its exceedingly gracious images. Julian was not a pagan; he was a Gnostic allured by the allegories of Greek polytheism, who had the misfortune to find the name of Jesus Christ less sonorous than that or Orpheus. The Emperor personally paid for the academical tastes of the philosopher and rhetorician, and after affording himself the spectacle and satisfaction of expiring like Epaminondas with the periods of Cato, he had in public opinion, already thoroughly Christianized, anathemas for his funeral oration and a scornful epithet for his ultimate celebrity.

Let us skip the little men and small matters of the Bas-Empire, and pass on to the Middle Ages. . . . Stay, take this book! Glance at the seventh page, then seat yourself on the mantle I am spreading, and let each of us cover our eyes with one of its corners. . . . Your head swims, does it not, and the earth seems to fly beneath your feet? Hold tightly, and do not look around. . . . The vertigo ceases; we are here. Stand up and open your eyes but take care before all things to make no Christian sign and to pronounce no Christian words. We are in a landscape of Salvatore Rosa, a troubled wilderness which seems resting after a storm; there is no moon in the sky, but you can distinguish little stars gleaming in the brushwood, and you can hear about you the slow flight of great birds, who seem to whisper strange oracles as they pass. Let us approach silently that cross-road among the rocks. A harsh, funereal trumpet winds suddenly, and black torches flare up on every side. A tumultuous throng is surging around a vacant throne; all look and wait. Suddenly they cast themselves on the ground. A goat-headed prince bounds forward among them; he ascends the throne, turns, and by assuming a stooping posture, presents to the assembly a human face, which, carrying black torches, everyone comes forward to salute and to kiss. With a hoarse laugh he recovers an upright position, and then distributes gold, secret instructions, occult medicines, and poisons to his faithful bondsmen. Meanwhile, fires are lighted of fern and alder, piled over with human bones and the fat of executed criminals. Druidesses crowned with wild parsley and vervain immolate unbaptized children with golden knives and prepare horrible love-feasts. Tables are spread, masked men seat themselves by half-nude females, and a Bacchanalian orgie begins; there is nothing missing but salt, the symbol of wisdom and immortality. Wine flows in streams, leaving stains like blood; obscene talk and fond caresses begin, and presently the whole assembly is drunk with wine, with pleasure, with crime, and singing. They rise, a disordered throng, and hasten to form infernal dances. . . . Then come all legendary monsters, all phantoms of nightmare; enormous toads play inverted flutes and blow with their paws on their flanks; limping scarab mingle in the dance; crabs play the castanets; crocodiles beat time on their scales; elephants and mammoths appear habited like Cupids and foot it in the ring; Finally, the giddy circles break up and scatter on all sides. . . . Every yelling dancer drags away a disheveled female. . . . Lamps and candles formed of human fat go out smoking in the darkness. . . . Cries are heard here and there, mingled with peals of laughter, blasphemies and rattlings of the throat. Come, rouse yourself, do not make the sign of the cross! See, I have brought you home; you are in your own bed, somewhat worn-out, possibly a trifle shattered, by your night's journey and dissipation; but you have witnessed something of which everyone talks without knowledge; you have been initiated into secrets no less terrible than the grotto of Triphonius; you have been present at the Sabbath. It remains for you now to preserve your reason, to have a wholesome dread of the law, and to keep at a respectful distance from the Church and her faggots.

Would you care, as a change, to behold something less fantastic, more real, and also more truly terrible? You shall assist at the execution of Jacques de Molay and his accomplices and his brethren in martyrdom. . . . Do not, however, be misled, confuse not the guilty and the innocent! Did the Templars really adore Baphomet? Did they offer a shameful salutation to the buttocks of the goat of Mendes? What was actually this secret and potent association which imperiled Church and State, and was thus destroyed unheard? Judge nothing lightly; they are guilty of a great crime; they have allowed the sanctuary of antique initiation to be entered by the profane. By them for a second time have the fruits of the tree of knowledge of good and evil been gathered and shared, so that they might become the masters of the world. The sentence which condemns them has a higher and earlier origin than the tribunal of pope or king: "On the day that thou eatest thereof, thou shalt surely die," said God Himself, as we see in the book of Genesis.

What is taking place in the world, and why do priests and potentates tremble? What secret powers threaten tiaras and crowns? A few madmen are roaming from land to land, concealing, as they say, the philosophical stone under their ragged vesture. They can change earth into gold, and they are without food or lodging! Their brows are encircled by an aureole of glory and by a shadow of ignominy! One has discovered the universal science and goes vainly seeking death to escape the agonies of his triumph – he is the Majorcan Raymond Lully. Another heals imaginary diseases by fantastic remedies, giving a formal denial in advance to the proverb which enforces the futility of a cautery on a wooden leg – he is the marvelous Paracelsus, always drunk and always lucid, like the heroes of Rabelais. Here is William Postel writing naïvely to the fathers of the Council of Trent, informing them that he has discovered the absolute doctrine, hidden from the foundation of the world, and is longing to share it with them. The council does not concern itself with the maniac, does not condescend to condemn him, and proceeds to examine the weighty questions of efficacious grace and sufficing grace. He whom we see perishing poor and abandoned is Cornelius Agrippa, less of a magician than any, though the vulgar persist in regarding him as a more potent sorcerer than all because he was sometimes a cynic and mystifier. What secret do these men bear with them to their tomb? Why are they wondered at without being understood? Why are they condemned unheard? Why are they initiates of those terrific secret sciences of which the Church and society are afraid? Why are they acquainted with things of which others know nothing? Why do they conceal what all men burn to know? Why are they invested with a dread and unknown power? The occult sciences! Magic! These words will reveal all and give food for further thought! *De omni re scibili et quibusdam aliis.*

But what, as a fact, was this magic? What was the power of these men who were at once so proud and so persecuted? If they were really strong, why did they not overcome their enemies? But if they were weak and foolish, why did people honor them by fearing them? Does magic exist? Is there an occult knowledge which is truly a power, which works wonders fit to be compared with the miracles of authorized religions? To these two palmary questions we make answer by an affirmation and a book. The book shall justify the affirmation, and the affirmation is this. *Yes*, there existed in the past. And there exists in the present, a potent and real magic; *yes*, all that legends have said of it is true, but, in contrariety to what commonly happens, popular exaggerations are, in this case, not only beside but below the truth. There is indeed a formidable secret, the revelation of which has once already transformed the world, as testified in Egyptian religious tradition, symbolically summarized by Moses at the beginning of Genesis. This secret constitutes the fatal science of good and evil, and the consequence of its revelation is death. Moses depicts it under the figure of a tree which is *in the center* of the Terrestrial Paradise, is in proximity to the tree of life and has radical connection therewith; at the foot of this tree is the source of four mysterious rivers; it is guarded by the sword of fire and by the four figures of the Biblical sphinx, the Cherubim of Ezekiel. . . . Here I must pause, and I fear already that I have said too much. Yes, there is one sole, universal, and imperishable dogma, strong as the supreme reason; simple, like all that is great; intelligible, like all that is universally and absolutely true; and this dogma has been the parent of all others. Yes, there is a science which confers on man powers apparently superhuman; I find them enumerated as follows in a Hebrew manuscript of the sixteenth century: –

"These are the powers and privileges of the man who holds in his right hand the clavicles of Solomon, and in his left the branch of the blossoming almond. א *Aleph*. – He beholds God face to face, without dying, and converses familiarly with the seven genii who command the entire celestial army. ב *Beth*. – He is above all afflictions and all fears. ג *Ghimel*. – He reigns with all heaven and is served by all hell. ד *Daleth*. – He disposes of his own health and life and can equally influence that of others. ה *He*. – He can neither be surprised by misfortune, nor overwhelmed by disasters, nor conquered by his enemies. ו *Vau*. – He knows the reason of the past, present and future. ז *Dzain*. – He possesses the secret of the resurrection of the dead and the key of immortality.

"Such are the seven chief privileges, and those which rank next are as follows: –

"ס *Samech*. – To know at first sight the deep things of the souls of men and the mysteries of the hearts of women. ע *Gnain*. – To force nature to make him free at his pleasure. פ *Phe*. – To foresee all future events which do not depend on a superior free will, or an indiscernible cause. צ *Tsade*. – To give at once and to all the most efficacious consolations and the most wholesome counsels. ק *Coph*. – To triumph over adversities. ר *Resch*. – To conquer love and hate. ש *Schin*. – To have the secret of wealth, to be always its master and never its slave. To know how to enjoy even poverty and never become abject or miserable. ♩ *Tau*. – Let us add to these three septenaries that the wise man rules the elements, stills tempests, cures the diseased by his touch, and raises the dead!

"At the same time, there are certain things which have been dealed by Solomon with his triple seal. It is enough that the initiates know, and as for others, whether they deride, doubt, or believe, whether they threaten or fear, what matters it to science or to us?"

Such are actually the issues of occult philosophy, and we are in a position to withstand an accusation of insanity or a suspicion of imposture when we affirm that all these privileges are real. To demonstrate this is the sole end of our work on occult philosophy. The philosophical stone, the universal medicine, the transmutation of metals, the quadrature of the circle, the secret of perpetual motion, are thus neither mystifications of science nor dreams of madness. They are terms which must be understood in their veritable sense; they are expressions of the different applications of one same secret, the several characteristics of one same operation, which is defined in a more comprehensive manner under the name of the great work. Furthermore, there exists in nature a force which is immeasurably more powerful than steam, and by means of which a single man, who knows how to adapt and direct it, might upset and alter the face of the world. This force was known to the ancients; it consists in a universal agent having equilibrium for its supreme law, while its direction is concerned immediately with the great arcanum of transcendent magic. By the direction of this agent it is possible to change the very order of the seasons; to produce at night the phenomena of day; to correspond instantaneously between one extremity of the earth and the other; to see, like Apollonius, what is taking place on the other side of the world; to heal or injure at a distance; to give speech a universal success and reverberation. This agent, which barely manifests under the uncertain methods of Mesmer's followers, is precisely that which the adepts of the middle ages denominated the first matter of the great work. The Gnostics represented it as the fiery body of the Holy Spirit; it was the object of adoration in the secret rites of the Sabbath and the Temple, under the hieroglyphic figure of Baphomet as the Androgyne of Mendes. All this will be proved.

Such are the secrets of occult philosophy, such is magic in history; let us now glance at it as it appears in its books and its achievements, in its initiations and its rites. The key of all magical allegories is found in the tablets we have already mentioned, and these tablets we regard as the work of Hermes. About this book, which may be called the keystone of the whole edifice of occult science, are grouped innumerable legends which are either its partial translation or its commentary renewed endlessly under a thousand different forms. Sometimes these ingenious fables combine harmoniously into a great epic which characterizes an epoch, though how or why is not clear to the uninitiated. Thus, the fabulous history of the Golden Fleece both resumes and veils the Hermetic and magical doctrines of Orpheus, and if we recur only to the mysterious poetry of Greece, it is because the sanctuaries of Egypt and India to some extent dismay us by their resources, and leave our choice embarrassed in the midst of such abundant wealth. We are eager, moreover, to reach the Thebaïd at once, that dread synthesis of all doctrine, past, present, and future, that, so to speak, infinite fable, which comprehends, like the Deity of Orpheus, the two extremities of the cycle of human life. Extraordinary fact! The seven gates of Thebes, attacked and defended by seven chiefs who have sworn upon the blood of victims, possess the same significance as the seven seals of the sacred book interpreted by seven genii, and assailed by a monster with seven heads, after being opened by a living yet immolated lamb, in the allegorical work of St. John. The mysterious origin of Œdipus, found suspended from the tree of Cytheron like a bleeding fruit, recalls the symbols of Moses and the narratives of Genesis. He makes war upon his father, whom he slays without knowing – alarming prophecy of the blind emancipation of reason without science; he then meets with the sphinx – the sphinx, that symbol of symbols, the eternal enigma of the vulgar, the granite pedestal of the science of the sages, the voracious and silent monster whose invariable form expresses the one dogma of the great universal mystery. How is the tetrad changed into the duad and explained by the triad? In more common but more emblematic terms, what is that animal that in the morning has four feet, two at noon, and three in the evening? Philosophically speaking, how does the doctrine of elementary forces produce the dualism of Zoroaster, while it is summed by the triad of Pythagoras and Plato? What is the ultimate reason of allegories and numbers, the final message of all symbolisms? Œdipus replies with a simple and terrible word which destroys the sphinx and makes the diviner the King of Thebes; the answer to the enigma is Man! . . . . Unfortunate! He has seen too much, and yet with insufficient clearness; he must presently expiate his calamitous and imperfect clairvoyance by a voluntary blindness and then vanish in the midst of a storm, like all civilizations which may at any time divine the answer to the riddle of the sphinx without grasping its whole import and mystery. Everything is symbolical and transcendental in this titanic epic of human destinies. The two hostile brethren express the second part of the grand mystery divinely completed by the sacrifice of Antigone; then comes the last war; the brethren slay one another, Capaneus is destroyed by the lightning which he defies, Amphiaraüs is swallowed by the earth, and all these are so many allegories which, by their truth and their grandeur, astonish those who can penetrate their triple hieratic sense. Æschylus, annotated by Ballanche, gives only a weak notion concerning them, whatever the primeval sublimities of the Greek poet or the beauty of the French critic.

The secret book of antique initiation was not unknown to Homer, who outlines its plan and chief figures on the shield of Achilles, with minute precision. But the gracious fictions of Homer replaced speedily in the popular memory the simple and abstract truths of primeval revelation. Humanity clung to the form and allowed the idea to be forgotten; signs lost power in their multiplication; magic also at this period became corrupted and degenerated with the sorcerers of Thessaly into the most profane enchantments. The crime of Œdipus brought forth its deadly fruits, and the science of good and evil erected evil into a sacrilegious divinity. Men, weary of the light, took refuge in the shadow of bodily substance; the dream of the void, which is filled by God, soon appeared to be greater than God himself in their eyes, and thus hell was created.

When, in the course of this work, we make use of the consecrated terms God, Heaven, and Hell, let it be thoroughly understood, once for all, that our meaning is as far removed from that which the profane attach to them as initiation is distant from vulgar thought. God, for us, is the AZOT of the sages, the efficient and final principle of the great work.

Returning to the fable of Œdipus, the crime of the King of Thebes was that he failed to understand the sphinx, that he destroyed the scourge of Thebes without being pure enough to complete the expiation in the name of the people. The plague, in consequence, avenged speedily the death of the monster, and the King of Thebes, forced to abdicate, sacrificed himself to the terrible manes of the sphinx, more alive and voracious then ever when it had passed from the domain of form into that of idea. Œdipus divined what was man and he put out his own eyes because he did not see what was God. He divulged half of the great arcanum, and, to save his people, it was necessary for him to bear the remaining half of the terrible secret into exile and the tomb.

After the colossal fable of Œdipus we find the gracious poem of Psyche, which was certainly not invented by Apuleius. The great magical arcanum reappears here under the figure of a mysterious union between a god and a weak mortal abandoned alone and naked on a rock. Psyche must remain in ignorance of the secret of her ideal royalty, and if she beholds her husband she must lose him. Here Apuleius commentates and interprets Moises, but did not the Elohim of Israel and the gods of Apuleius both issue from the sanctuaries of Memphis and Thebes? Psyche is the sister of Eve, or, rather, is Eve spiritualized. Both desire to know and lose innocence for the honor of the ordeal. Both deserve to go down into hell, one to bring back the antique box of Pandora, the other to find and to crush the head of the old serpent, who is the symbol of time and of evil. Both are guilty of the crime which must be expiated by the Prometheus of ancient days and the Lucifer of the Christian legend, the one delivered, the other overcome, by Hercules and by the Savior. The great magical secret is, therefore, the lamp and the dagger of Psyche, the apple of Eve, the sacred fire of Prometheus, the burning scepter of Lucifer, but it is also the holy cross of the Redeemer. To be acquainted with it sufficiently to abuse or divulge it is to deserve all sufferings; to know it as one should know it, namely, to make use of and conceal it, is to be master of the absolute.

Everything is contained in a single word, which consists of four letters; it is the Tetragram of the Hebrews, the Azot of the alchemists, the Thot of the Bohemians, or the Taro of the Kabbalists. This word, expressed after so many manners, means God for the profane, man for the philosophers, and imparts to the adepts the final word of human sciences and the key of divine power; but he only can use it who understands the necessity of never revealing it. Had Œdipus, instead of killing the sphinx, overcome it, harnessed it to his chariot, and thus entered Thebes, he would have been king without incest, without misfortunes, and without exile. Had Psyche, by meekness and affection, persuaded Love to reveal himself, she would never have lost Love. Now, Love is one of the mythological images of the great secret and the great agent, because it at once expresses an action and a passion, a void and a plenitude, a shaft and a wound. The initiates will understand me, and, on account of the profane, I must not speak more clearly.

After the marvelous Golden Ass of Apuleius, we find no more magical epics. Science, conquered in Alexandria by the fanaticism of the murderers of Hypatia, became Christian, or, rather, concealed itself under Christian veils with Ammonius, Synesius, and the pseudonymous author of the books of Dionysius the Areopagite. In such times it was needful to excuse miracles by the garb of superstition and science by an unintelligible language. Hieroglyphic writing was revived; pantacles and characters were invented to summarize an entire doctrine by a sign, a whole sequence of tendencies and revelations in a word. What was the end of the aspirants to knowledge? They sought the secret of the great work, or the philosophical stone, or the perpetual motion, or the quadrature of the circle, or the universal medicine – formulas which often saved them from persecution and hatred by causing them to be taxed with madness, and all signifying one of the phases of the great magical secret, as we shall show later on. This absence of epics continues until our *Romance of the Rose*; but the rose-symbol, which expresses also the mysterious and magical sense of Dante's poem, is borrowed from the transcendent Kabbalah, and it is time that we should have recourse to this immense and concealed source of universal philosophy.

The Bible, with all its allegories, gives expression to the religious knowledge of the Hebrews in only an incomplete and veiled manner. The book which we have mentioned, the hieratic characters of which we shall explain subsequently, that book which William Postel names the *Genesis of Enoch*, certainly existed before Moses and the prophets, whose doctrine, fundamentally identical with that of the ancient Egyptians, had also its esotericism and its veils. When Moses spoke to the people, says the sacred book allegorically, he placed a veil over his face, and he removed it when addressing God; this accounts for the alleged Biblical absurdities which so exercised the satirical powers of Voltaire. The books were only written as memorials of tradition, and in symbols that were unintelligible for the profane. The Pentateuch and the poems of the prophets were, moreover, elementary works, alike in doctrine, ethics, and liturgy; the true secret and traditional philosophy was not committed to writing until a later period, and under veils even less transparent. Thus arose a second and unknown Bible, or rather one which was not comprehended by Christians, a storehouse, so they say, of monstrous absurdities, for, in this case, believers, confounded in the same ignorance, speak the language of skeptics; a monument, as we affirm, which comprises all that philosophical genius and religious genius have ever accomplished or imagined in the order of the sublime; a treasure encompassed by thorns; a diamond concealed in a rude and opaque stone: our readers will have already guessed that we refer to the Talmud. How strange is the destiny of the Jews, those scapegoats, martyrs, and saviors of the world, a people full of vitality, a bold and hardy race, which persecutions have always preserved intact, because it has not yet accomplished its mission! Do not our apostolical traditions declare that, after the decline of faith among the Gentiles, salvation shall again come forth out of the house of Jacob, and that then the crucified Jew who is adored by the Christians will give the empire of the world into the hands of God his Father?

On penetrating into the sanctuary of the Kabbalah, one is seized with admiration at the sight of a doctrine so logical, so simple, and, at the same time, so absolute. The essential union of ideas and signs; the consecration of the most fundamental realities by primitive characters; the trinity of words, letters, and numbers; a philosophy simple as the alphabet, profound and infinite as the Word; theorems more complete and luminous than those of Pythagoras; a theology which may be summed up on the fingers; an infinite which can be held in the hollow of an infant's hand; ten figures and twenty-two letters, a triangle, a square, and a circle; these are the entire elements of the Kabbalah. These are the component principles of the written Word, reflection of that spoken Word which created the world! All truly dogmatic religions have issues from the Kabbalah and return therein; whatsoever is grand or scientific in the religious dreams of all the illuminated, Jacob Boehme, Swedenborg, Saint Martin, &c., is borrowed from the Kabbalah; all masonic associations owe to it their secrets and their symbols. The Kabbalah alone consecrates the alliance of universal reason and the divine Word; it establishes, by the counterpoise of two forces apparently opposed, the eternal balance of being; it only reconciles reason with faith, power with liberty, science with mystery; it has the keys of the present, past, and future!

To become initiated into the Kabbalah, it is insufficient to read and to meditate upon the writings of Reuchlin, Galatinus, Kircher, or Picus de Mirandola; it is necessary to study and to understand the Hebrew writers in the collection of Pistorius, the Septer Jetzirah above all; it is necessary also to master the great book Zohar, read attentively in the collection of 1684, entitled *Kabbala Denudata*, the treatise of Kabbalistic Pneumatics, and that of the Revolution of Souls; and afterwards to enter boldly into the luminous darkness of the whole dogmatic and allegorical body of the Talmud. Then we shall be in a position to understand William Postel, and can admit secretly that apart from his very premature and over-generous dreams about the emancipation of women, this celebrated, learned, illuminated man could not have been so mad as is pretended by those who have not read him.

We have sketched rapidly the history of occult philosophy; we have indicated its sources and analyzed in a few words its principal books. This work refers only to the science, but magic, or, rather, magical power, is composed of two things, a science and a force; without the force the science is nothing, or, rather, it is a danger. To give knowledge to power alone, such is the supreme law of initiations. Hence did the Great Revealer say: "The kingdom of heaven suffereth violence, and the violent only shall carry it away." The door of truth is closed like the sanctuary of a virgin; he must be a man who would enter. All miracles are promised to faith, and what is faith except the audacity of a will which does not hesitate in the darkness, but advances towards the light in spite of all ordeals, and surmounting all obstacles? It is unnecessary to repeat here the history of ancient initiations; the more dangerous and terrible they were, the greater was their efficacy. Hence, in those days, the world had men to govern and instruct it. The sacerdotal art and the royal art consisted above all in ordeals of courage, discretion, and will. It was a novitiate similar to that of those priests who, under the name of Jesuits, are so unpopular at the present day, but would govern the world, notwithstanding, had they a truly wise and intelligent chief.

After passing our life in the search after the absolute in religion, science, and justice; after turning in the circle of Faust, we have reached the primal doctrine and the first book of humanity. There we pause, there we have discovered the secret of human omnipotence and indefinite progress, the key of all symbolisms, the first and final doctrine, and we have come to understand what was meant by that expression so often made use of in the Gospel – the Kingdom of God.

To provide a fixed point as a fulcrum for human activity is to solve the problem of Archimedes by realizing the application of his famous lever. This it is which was accomplished by the great initiators who have electrified the world, and they could not have done so except by means of the great and incommunicable secret. However, as a guarantee of its renewed youth, the symbolical phœnix never reappeared before the eyes of the world without having solemnly consumed the remains and evidences of his previous life. It is thus that Moses caused all those to perish in the desert who could have known Egypt and her mysteries; thus, at Ephesus, St. Paul burnt all books which treated of the occult sciences; thus, finally, the French Revolution, daughter of the great Johannite Orient and the ashes of the Templars, spoliated the churches and blasphemed the allegories of the divine cultus. But all doctrines and all revivals proscribe magic, and condemn its mysteries to the flames and to oblivion. The reason is that each cultus or philosophy which comes into the world is a Benjamin of humanity which lives by the death of its mother; it is because the symbolical serpent seems ever devouring its own tail; it is because, as essential condition of existence, a void is necessary to every plenitude, space for every dimension, an affirmation for each negation; it is the eternal realization of the phœnix allegory.

Two illustrious scholars have already preceded me along the path I am travelling, but they have, so to speak, spent the dark night therein. I refer to Volney and Dupuis, to Dupuis above all, whose immense erudition has produced only a negative work, for in the origin of all religions he has seen nothing but astronomy, taking thus the symbolic cycle for doctrine and the calendar for legend. He was deficient in one branch of knowledge, that of true magic, which comprises the secrets of the Kabbalah. Dupuis passed through the antique sanctuaries like the prophet Ezekiel over the plain strewn with bones, and only understood death, for want of that word which collects the virtue of the four winds, and can make a living people of all the vast ossuary, by crying to the ancient symbols: "Arise! Take up a new form and walk!" Hence the hour has come when we must have the boldness to attempt what no one has dared to perform previously. Like Julian, we would rebuild the temple, and in so doing we do not believe that we shall be belying a wisdom that we adore, which also Julian would himself have been worthy to adore, had the rancorous and fanatical doctors of his period permitted him to understand it. For us the temple has two pillars, on one of which Christianity has inscribed its name. We have, therefore, no wish to attack Christianity; far from it, we seek to explain and accomplish it. Intelligence and will have alternately exercised their power in the world; religion and philosophy are still at war in our own days, but they must end by agreeing. The provisional object of Christianity was to establish, by obedience and faith, a supernatural or religious equality among men, and to immobilize intelligence by faith, so as to provide a fulcrum for virtue which came for the destruction of the aristocracy of science, or, rather, to replace that aristocracy already destroyed. Philosophy, on the contrary, has labored to bring back men by liberty and reason to natural inequality, and to substitute astuteness for virtue by inaugurating the reign of industry. Neither of the two operations has proved complete and adequate, neither has brought men to perfection and felicity. What is now dreamed, almost without daring to hope for it, is an alliance between those two forces so long regarded as contrary, and there is good ground for desiring their union, for these two great powers of the human soul are no more opposed to one another than the sex of man is opposed to that of woman; undoubtedly they differ, but their apparently contrary dispositions come only from their aptitude to meet and unite.

"There is no less proposed, therefore, than a universal solution of all problems?"

No doubt, since we are concerned with explaining the philosophical stone, perpetual motion, the secret of the great work and of the universal medicine. We shall be accused of insanity, like the divine Paracelsus, or of charlatanism, like the great and unfortunate Agrippa. If the pyre of Urban Grandier be extinguished, the sullen proscriptions of silence and calumny remain. We do not brace but are resigned to them. We have not sought ourselves the publication of this book, and we believe that if the time be come to produce speech, it will be produced by us or by others. We shall therefore remain calm and wait.

Our work has two parts; in the one we establish the Kabbalistic and magical doctrine in its entirety; the other is consecrated to the cultus, that is, to ceremonial magic. The one is that which the ancient sages termed the clavicle, the other that which rural people still call the grimoire. The numbers and subjects of the chapters, which correspond in both parts, are in no sense arbitrary, and are all indicated in the great universal key, of which we give for the first time a complete and adequate explanation. Let this work now go its way where it will, and become what Providence determines; it is finished, and we believe it to be enduring, because it is strong, like all that is reasonable and conscientious.

Éliphas Lévi.

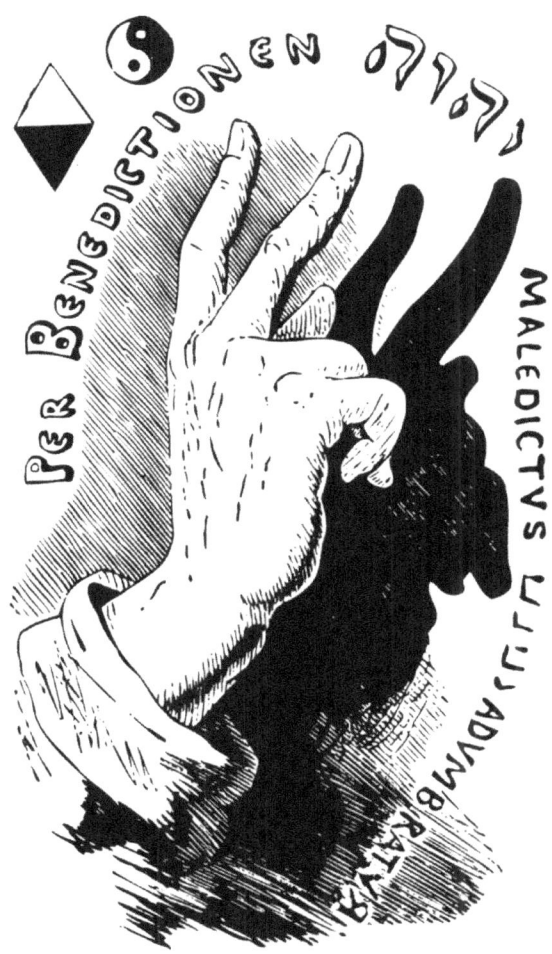

Transcendent Magic

I א A

THE CANDIDATE
DISCIPLINA ENSOPH KETER

WHEN a philosopher adopted as the basis for a new apocalypse of human wisdom the axiom: "I think, therefore I am," in a measure he unconsciously altered, from the standpoint of Christian Revelation, the old conception of the Supreme Being. I am that I am, said the Being of beings of Moses. I am he who thinks, says the man of Descartes, and to think being to speak inwardly, such a one may affirm like the God of St. John the Evangelist: I am he in whom and by whom the word manifests – *In principio erat verbum.*[8] Now, what is a principle? It is a groundwork of speech, it is a reason for the existence of the word. The essence of the word is in the principle; the principle is that which is; intelligence is a principle which speaks. What, further, is intellectual light? It is speech. What is revelation? It is also speech; being is the principle, speech is the means, and the plenitude or development and perfection of being is the end. To speak is to create. But to say: "I think, therefore I exist," is to argue from consequence to principle, and certain contradictions which have been adduced by a great writer, Lamennais, have abundantly proved the philosophical imperfection of this method. I am, therefore something exists – would appear to us a more primitive and simple foundation for experimental philosophy. I AM, THEREFORE BEING EXISTS. *Ego sum qui sum*[9] – such is the first revelation of God in man and of man in the world, while it is also the first axiom of occult philosophy. אהיה אשר אהיה. Being is being. Hence this philosophy, having that which is for its principle, can be in no sense hypothesis or guesswork.

---

[8] "In the beginning was the word." -pnw
[9] "I am who I am." -pnw

Mercurius Trismegistus begins his admirable symbol, known under the name of the Emerald Table, by this threefold affirmation: It is true, it is certain without error, it is of all truth. Thus, in physics, the true confirmed by experience; in philosophy, certitude purged from any alloy of error; in the domain of religion or the infinite, absolute truth indicated by analogy; such are the first necessities of true science, and magic only can impart these to its adepts.

But you, before all things, who are you, thus taking this work in your hands and proposing to read it? On the pediment of a temple consecrated by antiquity to the God of Light was an inscription of two words: "Know thyself." I impress the same counsel on every man when he seeks to approach science. Magic, which the men of old denominated the *sanctum regnum*, the holy kingdom, or kingdom of God, *regnum Dei*, exists only for kings and for priests. Are you priests? Are you kings? The priesthood of magic is not a vulgar priesthood, and its royalty enters not into competition with the princes of this world. The monarchs of science are the priests of truth, and their sovereignty is hidden from the multitude like their prayers and sacrifices. The kings of science are men who know the truth and the truth has made free, according to the specific promise given by the most mighty of all initiators.

The man who is enslaved by his passions or worldly prejudices can in no wise be initiated; he must alter or he will never attain; hence he cannot be an adept, for the word signifies a person who has attained by will and by work. The man who loves his own opinions and fears to part with them, who suspects new truths, who is unprepared to doubt everything rather than admit anything on chance, should close this book; for him it is useless and dangerous; he will fail to understand it, and it will trouble him, while if he should divine its meaning, there will be a still greater source of disquietude. If you hold by anything in the world more than by reason, truth and justice; if your will be uncertain and vacillating, either in good or evil; if logic alarm you, or the naked truth make you blush; if you are hurt when accepted errors are assailed; condemn this work straight away; do not read it; let it cease to exist for you; but at the same time do not cry it down as dangerous. The secrets which it records will be understood by an elect few and will be held back by those who understand them. Show light to the birds of the night-time, and you hide their light; it is the light which blinds them, and for them is darker than darkness. I shall therefore speak clearly and make known everything, with the firm conviction that initiates alone, or those who deserve initiation, will read all and understand in part.

There is a true and a false science, a divine magic and an infernal magic – in other words, one which is delusive and darksome; it is our task to reveal the one and to unveil the other, to distinguish the magician from the sorcerer, and the adept from the charlatan. The magician avails himself of a force which he knows, the sorcerer seeks to abuse a force which he does not understand. If it be possible in a scientific work to employ a term so vulgar and so discredited, then the devil gives himself to the magician and the sorcerer gives himself to the devil. The magician is the sovereign pontiff of nature, the sorcerer is her profaner only. The sorcerer bears the same relation to the magician that a superstitious and fanatical person bears to a truly religious man.

Before advancing further let us tersely define magic. Magic is the traditional science of the secrets of Nature which has been transmitted to us from the magi. By means of this science the adept becomes invested with a species of relative omnipotence and can operate superhumanly – that is, after a manner which transcends the normal possibility of men. Thereby many illustrious hierophants, such as Mercurius Trismegistus, Osiris, Orpheus, Apollonius of Tyana, and others whom it might be dangerous or unwise to name, came after their death to be adored and invoked as gods. Thereby others also, according to that ebb-and-flow of opinion which is responsible for the caprices of success – became emissaries of infernus or suspected adventurers, like the emperor Julian,

Transcendent Magic

Apuleius, the enchanter Merlin and that arch-sorcerer, as he was termed in his day, the illustrious and unfortunate Cornelius Agrippa.

To attain the *sanctum regnum*,[10] in other words, the knowledge and power of the magi, there are four indispensable conditions – an intelligence illuminated by study, an intrepidity which nothing can check, a will which nothing can break, and a discretion which nothing can corrupt and nothing intoxicate. TO KNOW, TO DARE, TO WILL, TO KEEP SILENCE – such are the four words of the magus, inscribed upon the four symbolical forms of the sphinx. These four words can be combined after four manners and explained four times by one another.[11]

On the first page of the Book of Hermes the adept is depicted with a large hat, which, if turned down, would conceal his entire head. One hand is extended towards heaven, which he seems to command with his rod, while the other is placed upon his breast; before him are the chief symbols or instruments of science, and he has others hidden in a juggler's wallet. His body and arms form the letter Aleph, the first of that alphabet which the Jews borrowed from the Egyptians; to this symbol we shall have occasion to recur later on.

The magus is truly that which the Hebrew Kabbalists call the *Microprosopus*, that is, the creator of the little world. The first of all magical sciences being the knowledge of one's self, so is one's own creation first of all works of science; it contains the others and is the principle of the great work. The term, however, requires explanation. Supreme reason being the sole invariable and consequently imperishable principle – that which we term death being change – hence the intelligence which cleaves closely to this principle and in a manner, identifies itself therewith, does hereby make itself unchangeable, and, as a result, immortal. To cleave invariably to reason, it will be understood that it is necessary to attain independence

---

[10] "Holy kingdom." -pnw
[11] See the Tarot cards.

of all those forces which by their fatal and inevitable movement produce the alternatives of life and death. To know how to suffer, to forbear, and to die – such are the first secrets which place us beyond reach of affliction, the desires of the flesh, and the fear of annihilation. The man who seeks and finds a glorious death has faith in immortality and universal humanity believes in it with him and for him, raising altars and statues to his memory in token of eternal life.

Man becomes king of the brutes only by subduing or taming them; otherwise he will be their victim or slave. Brutes are the type of our passions; they are the instinctive forces of nature. The world is a field of battle where liberty struggles with inertia by the opposition of active force. Physical laws are millstones; if you cannot be the miller you must be the grain. You are called to be king of the air, water, earth and fire; but to reign over these four living animals of symbolism, it is necessary to conquer and enchain them. He who aspires to be a sage and to know the great enigma of nature must be the heir and despoiler of the sphinx; his the human head in order to possess speech, his the eagle's wings in order to scale the heights, his the bull's flanks in order to furrow the depths, his the lion's talons to make a way on the right and the left, before and behind.

You, therefore, who seek initiation, are you learned as Faust? Are you insensible as Job? No, is it not so? But you may become equal to both if you will. Have you overcome the vortices of vague thoughts? Are you without indecision or capriciousness? Do you consent to pleasure only when you will, and do you wish for it only when you should? No, is it not so? Not invariably at least, but this may become so if you choose. The sphinx has not only a man's head, it has woman's breasts; do you know how to resist feminine charms? No, is it not so? And you laugh outright in replying, vaunting your moral weakness for the glorification of your physical and vital force. Be it so; I allow you to render this homage to the ass of Sterne or Apuleius. The ass has its merit, I agree; it was consecrated to Priapus as was the goat to the god of Mendes. But take it for what it is worth, and decide whether ass or man shall be master. He alone can possess truly the pleasure of love who has conquered the love of pleasure. To be able and to forbear is to be twice able. Woman enchains you by your desires; master your desires and you will enchain her. The greatest injury that can be inflicted on a man is to call him a coward. Now, what is a cowardly person? One who neglects his moral dignity in order to obey blindly the instincts of nature. As a fact, in the presence of danger it is natural to be afraid and seek flight; why, then, is it shameful? Because honor has erected it into a law that we must prefer our duty to our inclinations or fears. What is honor from this point of view? It is universal presentience of immortality and appreciation of the means which can lead to it. The last trophy which man can win from death is to triumph over the appetite for life, not by despair, but by a more exalted hope, which is contained in faith, for all that is noble and honest, by the undivided consent of the world. To learn self-conquest is therefore to learn life, and the austerities of stoicism were no vain parade of freedom! To yield to the forces of nature is to follow the stream of collective life, and to be the slave of secondary causes. To resist and subdue nature is to make for one's self a personal and imperishable life; it is to break free from the vicissitudes of life and death. Every man who is prepared to die rather than renounce truth and justice is most truly living, for immortality abides in his soul. To find or to form such men was the end of all ancient initiations. Pythagoras disciplined his pupils by silence and all kinds of self-denial; candidates in Egypt were tried by the four elements; and we know the self-inflicted austerities of fakirs and brahmans in India for attaining the kingdom of free will and divine independence. All macerations of asceticism are borrowed from the initiations of the ancient mysteries; they have ceased because those qualified for initiation, no longer finding initiators, and the leaders of conscience becoming in the lapse of time as uninstructed as the vulgar, the blind have grown weary of following the blind, and no one has cared to pass through ordeals the end of which was now only in doubt and despair; for the path of light was lost. To succeed in performing something we must know what it is proposed to do, or at least must have faith in someone who does know it. But shall I stake my life on a venture, or follow someone at chance who himself knows not where he is going?

We must not set out rashly along the path of the transcendent sciences, but, once started, we must reach the end or perish. To doubt is to become a fool; to pause is to fall; to recoil is to cast one's self into an abyss. You, therefore, who are undertaking the study of this book, if you persevere with it to the close and understand it, it will make you either a monarch or madman. Do what you will with the volume, you will be unable to despise or to forget it. If you are pure, it will be your light; if strong, your arm; if holy, your religion; if wise, the rule of your wisdom. But if you are wicked, for you it will be an infernal torch; it will lacerate your breast like a poniard; it will rankle in your memory like a remorse; it will people your imagination with chimeras, and will drive you through folly to despair. You will endeavor to laugh at it, and will only gnash your teeth; this book will be the file in the fable which the serpent tried to bite, but it destroyed all his teeth.

Let us now enter on the series of initiations. I have said that revelation is the word. As a fact, the word, or speech, is the veil of being and the characteristic sign of life. Every form is the veil of a word, because the idea which is the mother of the word is the sole reason for the existence of forms. Every figure is a character, every character derives from and returns into a word. For this reason the ancient sages, of whom Trismegistus is the organ, formulated their sole dogma in these terms: – "That which is above is like that which is below, and that which is below is like that which is above." In other words, the form is proportional to the idea; the shadow is the measure of the body calculated with its relation to the luminous ray; the scabbard is as deep as the sword is long; the negation is in proportion to the contrary affirmation; production is equal to destruction in the movement which preserves life; and there is no point in infinite extension which may not be regarded as the center of a circle having an expanding circumference indefinitely receding into space. Every individuality is, therefore, indefinitely perfectible, since the moral order is analogous to the physical, and since we cannot conceive any point as unable to dilate, increase and radiate in a philosophically infinite circle. What can be affirmed of the soul in its totality may be affirmed of each faculty of the soul. The intelligence and will of man are instruments of incalculable power and capacity. But intelligence and will possess as their help-mate and instrument a faculty which is too imperfectly known, the omnipotence of which belongs exclusively to the domain of magic. I speak of the imagination, which the Kabbalists term the *Diaphane*, or the *Translucid*. Imagination, in effect, is like the soul's eye; therein forms are outlined and preserved; thereby we behold the reflections of the invisible world; it is the glass of visions and the apparatus of magical life; by its intervention we heal diseases, modify the seasons, drive death away from the living, and raise the dead to life, because it is the imagination which exalts will and gives it a hold over the universal agent. Imagination determines the shape of the child in its mother's womb and decides the destiny of men; it lends wings to contagion, and directs the weapons of warfare. Are you exposed in battle? Believe yourself to be invulnerable, like Achilles, and you will be so, says Paracelsus. Fear attracts bullets, but they are repelled by courage. It is well known that persons with amputated limbs feel pain in the very members which they possess no longer. Paracelsus operated upon living blood by medicating the product of a bleeding; he cured headache at a distance by treating hair cut from the patient. By the science of the imaginary unity and solidarity between all parts of the body, he anticipated and outstripped all the theories, or rather experiences, of our most celebrated magnetizers. Hence his cures were miraculous, and to his name of Philip Theophrastus Bombast, he deserved the addition of Aureolus Paracelsus, with the further epithet of divine!

Imagination is the instrument of *the adaptation of the word*. Imagination applied to reason is genius. Reason is one, as genius is one, in the multiplicity of its works. There is one principle, there is one truth, there is one reason, there is one absolute and universal philosophy. Whatsoever is subsists in unity considered as beginning, and returns into unity considered as end. One is in one; that is to say, all is in all. Unity is the principle of numbers; it is also the principle of motion, and, consequently of life. The entire human body is summed up in the unity of a single organ, which is the brain. All religions are summed up in the unity of a single dogma, which is the affirmation of being and its equality with itself, which constitutes its mathematical value. There is only one dogma in magic, and it is this: – The visible is the manifestation of the invisible, or, in other terms, the perfect word, in things appreciable and visible, bears an exact proportion to the things which are inappreciable by our senses and unseen by our eyes. The magus raises one hand towards heaven and points down with the other to earth, and he says: "Above, immensity: Below immensity still! Immensity equals immensity." – This is true in things seen as in things unseen.

The first letter in the alphabet of the sacred language, Aleph, א, represents a man extending one hand towards heaven and the other to earth. It is an expression of the active principle in everything; it is creation in heaven corresponding to the omnipotence of the word below. This letter is a pantacle in itself, that is, a character expressing the universal science. It is supplementary to the sacred signs of the Macrocosm and the Microcosm; it explains the masonic double-triangle and five-pointed blazing star; for the word is one and revelation is one. By endowing man with reason, God gave him speech; and revelation, manifold in its forms but one in its principle, consists entirely in the universal word, the interpreter of the absolute reason. This is the significance of that term so much misconstrued, *catholicity*, which, in modern hieratic language, means *infallibility*. The universal in reason is the absolute, and the absolute is the infallible. If absolute reason impelled universal society to believe irresistibly the utterance of a child, that child would be infallible by the ordination of God and of all humanity. Faith is nothing else but reasonable confidence in this unity of reason and in this universality of the word. To believe is to place confidence in that which we as yet do not know when reason assures us beforehand of ultimately knowing or at least recognizing it. Absurd are the so-called philosophers who cry, "I will never believe in a thing which I do not know!" Shallow reasoners! If you knew, would you need to believe?

But must I believe on chance, and apart from reason? Certainly not. Blind and haphazard belief is superstition and folly. We may believe in causes which reason compels us to admit on the evidence of effects known and appreciated by science. Science! Great word and great problem! What is science? We shall answer in the second chapter of this book.

II ב B

THE PILLARS OF THE TEMPLE

CHOCHMAH DOMUS GNOSIS

SCIENCE is the absolute and complete possession of truth. Hence have the sages of all the centuries trembled before such an absolute and terrible word; they have hesitated to arrogate to themselves the first privilege of divinity by assuming the possession of science, and have been contented, instead of the verb *to know*, with that which expresses cognizance, while, instead place of the word *science*, they have adopted that of *gnosis*, which represents simply the notion of learning by intuition. What, in fact, does man know? Nothing, and at the same time he is allowed to ignore nothing. Devoid of knowledge, he is called upon to know all. Now, knowledge supposes the duad – a being who knows and an object known. The duad is the generator of society and of law; it is also the number of the gnosis. The duad is unity multiplying itself in order to create, and hence in sacred symbolism Eve issues from the inmost bosom of Adam. Adam is the human tetragram, summed up in the mysterious Jod, type of the Kabbalistic phallus. By adding to this Jod the triadic name of Eve, the name of Jehova is formed, which is eminently the kabbalistic and magical word, יהוה , which the high-priest in the temple pronounced Jodcheva. So unity complete in the fruitfulness of the triad forms therewith the tetrad, which is the key of all numbers, of all movements and of all forms. By a revolution about its own center, the square produces a circle equal to itself, and this is the quadrature of the circle, the circular movement of

Transcendent Magic

four equal angles around the same point.

"That which is above equals that which is below," says Hermes. Here then is the duad serving as the measure of unity, and the relation of equality between above and below forms with these the triad. The creative principle is the ideal phallus; the created principle is the formal cteïs. The insertion of the vertical phallus into the horizontal cteïs forms the stauros of the Gnostics, or the philosophical cross of masons. Thus, the intersection of two produces four, which, by its movement, defines the circle with all degrees thereof.

א is man; ב is woman; 1 is the principle; 2 is the word; A is the active; B is the passive; the monad is Bohas; the duad is Jakin. In the trigrams of Fohi, unity is the yang and the duad is the yin.

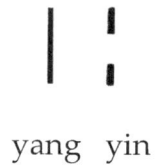

yang yin

Bohas and Jakin are the names of the two symbolical pillars without the chief door of Solomon's Kabbalistic temple. In the Kabbalah these Pillars explain all mysteries of antagonism, whether natural, political or religious, and they explain also the procreative struggle between man and woman, for, according to the law of Nature, the woman must resist the man, and he must entice or overcome her. The active principle seeks the passive principle, the *plenum* desires the void, the serpent's jaw attracts the serpent's tail, and in turning upon himself, he, at the same time, flies and pursues himself. Woman is the creation of man, and universal creation is the bride of the First Principle.

When the Supreme Being became a creator, he erected a jod or a phallus, and to provide a place in the fulness of the uncreated light, it was necessary to hollow out a cteïs or trench of shadow equivalent to the dimension determined by his creative desire, and attributed by him to the ideal jod of the radiating light. Such is the mysterious language of the Kabbalists in the Talmud, and on account of vulgar ignorance and malignity, it is impossible for us to explain or simplify it further. What then, is the creation? It is the mansion of the creative Word. What is the cteïs? It is the mansion of the phallus. What is the nature of the active principle? To diffuse. What is that of the passive? To gather in and to fructify. What is man? He who initiates, who bruises, who labors, who sows. What is woman? She who forms, reunites, irrigates and harvests. Man wages war, woman brings peace about; man destroys to create, woman builds up to preserve; man is revolution, woman is conciliation; man is the father of Cain, woman the mother of Abel. What, moreover, is wisdom? It is the agreement and union of two principles, the mildness of Abel directing the activity of Cain, man guided by the sweet inspirations of woman, debauchery conquered by lawful marriage, revolutionary energy softened and subdued by the gentleness of order and peace, pride subjugated by love, science acknowledging the inspirations of faith. Then human science becomes wise, and submits itself to the infallibility of universal reason, instructed by love or universal charity. Then it can assume the name of gnosis, because it knows at least that as yet it cannot boast of knowing perfectly.

The monad can only manifest by the duad; unity itself and the notion of unity at once constitute two. The unity of the Macrocosm reveals itself by the two opposite points of two triangles. Human unity fulfils itself to right and left. Primitive man is androgynous. All organs of the human body are disposed in pairs, excepting the nose, the tongue, the umbilicus and the Kabbalistic jod. Divinity, one in its essence, has two essential conditions as the fundamental grounds of its being – necessity and liberty. The laws of supreme reason necessitate and rule liberty in God, who is of necessity wise and reasonable.

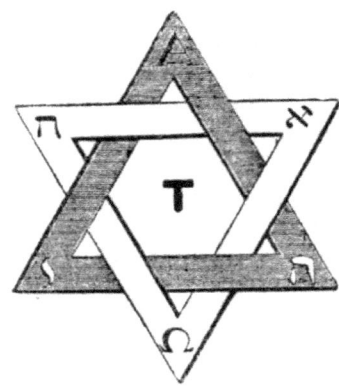

To make light visible God has merely hypothecated the shadow. To manifest the truth he has permitted the possibility of doubt. The shadow bodies forth the light, and the possibility of error is requisite for the temporal manifestation of truth. If the buckler of Satan did not intercept the spear of Michael, the might of the angel would be lost in the void or manifested by infinite destruction launched below from above. Did not the heel of Michael restrain Satan in his ascent, Satan would dethrone God, or rather he would lose himself in the abysses of the altitude. Hence Satan is needful to Michael as the pedestal to the statue, and Michael is necessary to Satan as the brake to the locomotive. In analogical and universal dynamics, one leans only on that which resists. Furthermore, the universe is balanced by two forces which maintain it in equilibrium, the force which attracts and that which repels. They exist alike in physics, in philosophy and in religion; in physics they produce equilibrium, in philosophy criticism, in religion progressive revelation. The ancients represented this mystery by the conflict between Eros and Anteros, the struggle between Jacob and the angel, and by the equilibrium of the golden mountain, which gods on the one side and demons on the other hold bound by the symbolic serpent of India. It is typified also by the caduceus of Hermanubis, by the two cherubim of the ark, by the twofold sphinx of the chariot of Osiris, and by the two seraphim, respectively black and white. Its scientific reality is demonstrated by the phenomena of polarity, and by the universal law of sympathies and antipathies.

The undiscerning disciples of Zoroaster divided the duad without referring it to unity, thus separating the pillars of the temple and endeavoring to halve God. Conceive the absolute as two, and you must immediately conceive it as three to recover the unity principle. For this reason, the material elements, analogous to the divine elements, are understood firstly as four, explained as two and exist ultimately as three.

Revelation is the duad; every word is double, and supposes two. The ethic which results from revelation is founded on antagonism, which results from the duad. Spirit and form attract and repel one another, like sign and idea, fiction and truth. Supreme Reason necessitates dogma when communicating to finite intelligences, and dogma, by its passage from the domain of ideas to that of forms, participates in two worlds and has inevitably two senses speaking in succession or simultaneously, that is, to the spirit and the flesh. So are there two forces in the moral region, one which assaults and one which curbs and expiates. They are represented in the mythos of Genesis by the typical personalities of Cain and Abel. Abel oppresses Cain by reason of his moral superiority; Cain to get free immortalizes his brother by slaying him, and becomes the victim of his own crime. Cain could not suffer the life of Abel, and the blood of Abel suffers not the sleep of Cain. In the Gospel the type of Cain is replaced by that of the Prodigal Son, whom his father fully forgives because he returns after having endured much.

There is mercy and there is justice in God; to the just He dispenses justice and to sinners mercy. In the soul of the world, which is the universal agent, there is a current of love and a current of wrath. This ambient and all-penetrating fluid; this ray loosened from the sun's splendor, and fixed by the weight of the atmosphere and the power of central attraction; this body of the Holy Spirit, which we term the universal agent, while it was typified by the ancients under the symbol of a serpent devouring his tail; this electro-magnetic ether, this vital and luminous caloric, is depicted in archaic monuments by the girdle of Isis, twice-folded in a love-knot round two poles, as well as by the serpent devouring his own tail, emblematic of prudence and of Saturn. Motion and life consist in the extreme tension of two forces. "I would thou wert cold or hot," said the Master. As a fact, a great sinner is more really alive than is a tepid, effeminate man, and the fulness of his return to virtue will be in proportion to the extent of his errors. She who is destined to crush the serpent's head is intelligence, which ever rises above the stream of blind forces. The Kabbalists call her the virgin of the sea, whose dripping feet the infernal dragon, stupefied by delight, crawls forward to lick with his fiery tongues. These are the hieratic mysteries of the duad. But there is one, and the last of all, which must not be made known, the reason, according to Hermes Trismegistus, being the malcomprehension of the vulgar, who would ascribe to the necessities of science the immoral aspect of blind fatality. "By the fear of the unknown must the crowd be restrained," he observes in another place; and Christ also said: "Cast not your pearls before swine, lest, trampling them under their feet, they turn and rend you." The Tree of the Knowledge of Good and Evil, of which the fruits are death, is the type of this hieratic secret of the duad, which would be only misconstrued if divulged, and would lead commonly to the unholy denial of free will, which is the principle of moral life. It is hence in the essence of things that the revelation of this secret means death, and it is not at the same time the great secret of magic; but the arcanum of the duad leads up to that of the tetrad, or more correctly proceeds therefrom, and is resolved by the triad, which contains the word of the enigma propounded by the sphinx, as it was required to have been found in order to save the life, atone for the unconscious crime, and establish the Kingdom of Œdipus.

In the hieroglyphic work of Hermes, the Tarot, called also the Book of Thoth, the duad is represented either by the horns of Isis, having her head veiled and an open book partially concealed under her mantle, or otherwise by a sovereign lady, Juno, the Greek goddess, with one hand uplifted towards heaven and the other pointed to earth, as if formulating by this gesture the one and twofold dogma which is the foundation of magic and begins the marvelous symbols of the Emerald Table of Hermes. In the Apocalypse of St. John there is a reference to two witnesses or martyrs on whom prophetic tradition confers the names of Elias and Enoch-Elias, man of faith, enthusiasm, miracle; Enoch one with him who is called Hermes by the Egyptians, honored by the Phœnicians as Cadmus, author of the sacred alphabet, and the universal key to the initiations of the Logos, father of the Kabbalah, he who, according to sacred allegories, did not die like other men, but was translated to heaven, to return at the end of time. Much the same statement is made of St. John himself, who recovered and explained in his Apocalypse the symbolism of the word of Enoch. This resurrection of St. John and Enoch, expected at the close of the ages of ignorance, will be the renovation of their doctrine by the comprehension of the Kabbalistic keys which unlock the temple of unity and universal philosophy, too long occult, and reserved solely for the elect, who perish at the hands of the world.

But we have said that the reproduction of the monad by the duad leads of necessity to the conception and dogma of the triad, so we come now to this great number, which is the fullness and perfect word of unity.

3 ג C

THE TRIANGLE OF SOLOMON

PLENITUDO VOCIS BINAH PHYSIS

THE perfect word is the triad, because it supposes an intelligent principle, a speaking principle and a principle spoken. The absolute, revealing itself by speech, endows this speech with a sense equivalent to itself, and in the understanding thereof creates itself a third time. Thus, also, the sun manifests by its light, and proves or makes this manifestation efficacious by heat.

The triad is delineated in space by the heavenly zenith, the infinite height, connected with East and West by two straight diverging lines. With this visible triangle reason compares another which is invisible, but is assumed to be equal in dimension; the abyss is its apex and its reversed base is parallel to the horizontal line stretching from east to west. These two triangles, combined in a single figure, which is the six-pointed star, form the sacred symbol of Solomon's seal, the resplendent star of the Macrocosm. The notion of the infinite and the absolute is expressed by this sign, which is the grand pantacle – that is to say, the most simple and complete abridgement of the science of all things.

Grammar itself attributes three persons to the verb. The first is that which speaks, the second that which is spoken to, and the third the object. In creating, the Infinite Prince speaks to himself of himself. Such is the explanation of the triad and the origin of the dogma of the Trinity. The magical dogma is also one in three and three in one. That which is above is like or equal to that which is below. Thus, two things which resemble one another and the word which signifies their resemblance make three. The triad is the universal dogma. In Magic – principle, realization, adaptation; in alchemy – azoth, incorporation, transmutation; in theology – God, incarnation, redemption; in the human soul – thought, love and action; in the family – father, mother and child. The triad is the end and supreme expression of love; we seek one another as two only to become three.

There are three intelligible worlds which correspond one with another by hierarchic analogy; the natural or physical, the spiritual or metaphysical, and the divine or religious worlds. From this principle follows the hierarchy of spirits, divided into three orders, and again subdivided by the triad in each of these three orders.

All these revelations are logical deductions from the first mathematical notions of being and number. Unity must multiply itself in order to become active. An indivisible, motionless and sterile principle would be unity dead and incomprehensible. Were God only one He would never be creator or father. Were he two there would be antagonism or division in the infinite, which would mean the division also or death of all possible things. He is therefore three for the creation by Himself and in His image of the infinite multitude of beings and numbers. So is He truly one in Himself and triple in our conception, which also leads us to behold Him as triple in Himself and one in our intelligence and our love. This is a mystery for the faithful and a logical necessity for the initiate into absolute and real sciences.

The Word manifested by life is realization or incarnation. The life of the Word accomplishing its cyclic movement is adaptation or redemption. This triple dogma was known in all sanctuaries illuminated by the tradition of the sages. Do you wish to ascertain which is the true religion? Seek that which realizes most in the Divine Order, which humanizes God and makes man divine, which preserves the triadic dogma intact, which clothes the Word with flesh by making God manifest to the hands and eyes of the most ignorant, which finally is by its doctrine suitable to all and can adapt itself to all – the religion which is hierarchic and cyclic, having allegories and images for children, an exalted philosophy for grown men, sublime hopes and sweet consolations for the old.

The primeval sages, when seeking the First of Causes, behold good and evil in the world; they considered the shadow and the light; they compared winter with spring, age with youth, life with death, and their conclusion was this: The First Cause is beneficent and severe; it gives and takes away life. Then are there two contrary principles, the one good and the other evil, exclaimed the disciples of Manes. No, the two principles of universal equilibrium are not contrary, although contrasted in appearance, for a singular wisdom opposes one to another. Good is on the right, evil on the left, but the supreme excellence is above both, applying evil to the victory of good and good to the amendment of evil.

The principle of harmony is in unity, and it is this which imparts such power to the uneven number in magic. Now, the most perfect of the odd numbers is three, because it is the trilogy of unity. In the trigrams of Fohi, the superior triad is composed of three yang, or masculine figures, because nothing passive can be admitted into the idea of God, considered as the principle of production in the three worlds. For the same reason, the Christian trinity by no means permits the personification of the mother, who is implicitly included in that of the Son. For the same reason, also, it is contrary to the laws of hieratic and orthodox symbology to personify the Holy Ghost under the form of a woman. Woman comes forth from man as nature comes forth from God; so Christ ascends Himself to heaven, and assumes the Virgin Mother: we speak of the ascension of the Savior, and the assumption of the Mother of God. God, considered as Father, has nature for his daughter; as Son, He has the Virgin for His mother and the Church for His bride; as Holy Spirit, He regenerates and fructifies humanity. Hence, in the trigrams of Fohi, the three inferior *yin* correspond to the three superior *yang*, for these trigrams constitute a pantacle like that of the two triangles of Solomon, but with a triadic interpretation of the six points of the blazing star.

Dogma is only divine inasmuch as it is truly human – that is to say, in so far as it sums up the highest reason of humanity; so also the Master, whom we term the Man-God, called Himself the Son of Man. Revelation is the expression of belief accepted and formulated by universal reason in the human word, on which account it is said that the divinity is human and the humanity divine in the Man-God. We affirm all this philosophically, not theologically, without infringing in any way on the teaching of the Church, which condemns, and must always condemn, magic. Paracelsus and Agrippa did not set up altar against altar, but bowed to the ruling religion of their time; to the elect of science, the things of science; to the faithful, the things of faith.

In his hymn to the royal Sun, the Emperor Julian gives a theory of the triad which is almost identical with that of the illuminated Swedenborg. The sun of the divine world is the infinite, spiritual and uncreated light, which is verbalized, so to speak, in the philosophical world, and becomes the fountain of souls and of truth; then it incorporates and becomes visible light in the sun of the third world, the central sun of our suns, of which the fixed stars are the ever-living sparks. The Kabbalists compare the spirit to a substance which remains fluid in the divine medium and under the influence of the essential light, its exterior, however, becoming solidified, like wax when exposed to the air in the colder realms of reasoning or of visible forms. These shells, envelopes petrified or carnified, were such an expression possible, are the source of errors or of evil, which connect with the heaviness and hardness of animal envelopes. In the book "Zohar," and in that of the "Revolution of Souls," perverse spirits or evil demons are never named otherwise than as shells – *cortices*. The cortices of the world of spirits are transparent, while those of the material world are opaque. Bodies are only temporary shells, whence souls have to be liberated; but those which in this life obey the body compose for themselves an interior body or fluidic shell, which, after death, becomes their prison-house and torment, until the time arrives when they succeed in dissolving it in the warmth of the divine light, towards which, however, the burden of their grossness hinders them from ascending. Indeed, they can do so only after infinite struggles, and by the mediation of the just, who stretch forth their hands towards them. During the whole period of the process they are devoured by the interior activity of the captive spirit, as in a burning furnace. Those who attain the pyre of expiation burn themselves thereon, like Hercules upon Mount Etna, and so are delivered from their sufferings; but the courage of the majority fails before this ordeal, which seems to them a second death more appalling than the first, and so they remain in hell, which is rightly and actually, eternal; but souls are never precipitated, nor are they ever retained despite themselves.

The three worlds correspond together by means of the thirty-two paths of light which are the steps of the sacred ladder; every true thought corresponds to a divine grace in heaven and a good work on earth; every grace of God manifests a truth, and produces one or many acts; reciprocally, every act affects a truth or falsehood in the heavens, a grace or a punishment. When a man pronounces the Tetragram – say the Kabbalists – the nine celestial realms sustain a shock, and then all spirits cry out one upon another: "Who is it thus disturbing the kingdom of heaven?" Then does the earth communicate unto the first heaven the sins of that rash being who takes the Eternal Name in vain, and the accusing word is transmitted from circle to circle, from star to star, and from hierarchy to hierarchy.

Every speech possesses three senses, every act has a triple bearing, every form a triple idea, for the absolute corresponds from world to world by its forms. Every determination of human will modifies nature, concerns philosophy and is written in heaven. There are therefore two fatalities, one resulting from the Uncreated Will in harmony with its proper wisdom, the other from created wills in accordance with the necessity of secondary causes in their correspondence with the First Cause. There is hence nothing indifferent in life, and our seeming most simple resolutions do often determine an incalculable series of benefits or evils, above all in the affinities of our diaphane with the great magical agent, as we shall explain elsewhere.

The triad, being the fundamental principle of the whole Kabbalah, or sacred tradition of our fathers, was necessarily the fundamental dogma of Christianity, the apparent dualism of which it explains by the intervention of a harmonious and all-powerful unity. Christ did not put his teaching into writing, and only revealed it in secret to his favored disciple, the one kabbalist, and he a great kabbalist, among the apostles. So is the apocalypse the book of the gnosis or secret doctrine of the first Christians, the key of which doctrine is indicated by an occult versicle of the Lord's Prayer, which the Vulgate leaves untranslated, while in the Greek rite, which preserves the traditions of St. John, the priests only are permitted to pronounce it. This versicle, completely kabbalistic, is found in the Greek text of the Gospel according to St. Matthew, and in several Hebrew copies. The sacred word Malchuth substituted for Kether, which is its kabbalistic correspondent, and the balance of Geburah and Chesed, repeating itself in the circles or heavens called eons by the Gnostics, provide the keystone of the whole Christian Temple in this occult versicle. It has been retained by Protestants in their New Testament, without their recovering its lofty and wonderful meaning, which would have unveiled to them all the Mysteries of the Apocalypse. But it is a tradition in the Church that the manifestation of these mysteries is held over to the last times.

Malchuth, based upon Geburah and Chesed, is the Temple of Solomon having Jakin and Boaz for its Pillars; it is the adamic doctrine founded, for the one part, on the resignation of Abel and, for the other, on the labors and self-reproach of Cain; it is the equilibrium of being established on necessity and liberty, stability and motion; it is the demonstration of the universal lever sought in vain by Archimedes. A scholar whose whole talents were employed in being obscure, who died without seeking to be understood, resolved this supreme equation, discovered by him in the Kabbalah, and was in dread of its source transpiring if he expressed himself more clearly. We have seen one of his disciples and admirers most indignant, perhaps in good faith, at the suggestion that his master was a Kabbalist, but we can state notwithstanding, to the glory of the same learned man, that his researches have appreciably shortened our work on the occult sciences, and that the key of the transcendent Kabbalah above all, indicated in the arcane versicle recently cited, has been applied skillfully to an absolute reform of all sciences in the books of Hœné Wronski.

The secret virtue of the gospels is therefore contained in three words, and these three words have established three dogmas and three hierarchies. All science reposes upon three principles, as the syllogism upon three terms. There are also three distinct classes, or three original and natural ranks, among men, who are called to advance from the lower to the higher. The Jews term these three series or degrees in the progress of spirits, Asiah, Jetzirah and Briah. The Gnostics, who were Christian Kabalists, called them Hylè, Psyche and Gnosis; by the Jews the supreme circle was named Atziluth, and by the Gnostics Pleroma. In the Tetragram, the triad, taken at the beginning of the word, expresses the divine copulation; taken at the end, it expressed the female and maternity. Eve has a name of three letters, but the primitive Adam is signified simply by the letter Jod, whence Jehovah should be pronounced Jeva, and this point leads us to the great and supreme mystery of magic, embodied in the tetrad.

4 ד D

THE TETRAGRAM

GEBURAH CHESED   PORTA LIBRORUM   ELEMENTA

IN nature there are two forces producing equilibrium, and these three constitute a single law. Here, then, is the triad resumed in unity, and by adding the conception of unity to that of the triad we are bought to the tetrad, the first square and perfect number, the source of all numerical combinations and the principle of all forms. Affirmation, negation, discussion, solution, such are the four philosophical operations of the human mind. Discussion conciliates negation with affirmation by rendering them necessary to each other. In the same way, the philosophical triad, emanating from the antagonism of the duad, is completed by the tetrad, the four-square ground of all truth. According to the consecrated dogma, there are three persons in God, and these three constitute only one Deity. Three and one provide the conception of four, because unity is required to explain the three. Hence, in almost all languages, the name of God consists of four letters, and in Hebrew these four are really three, one of them being repeated twice, that which expresses the Word and the creation of the Word.

Two affirmations make two corresponding denials either possible or necessary. Being is declared, nothing is not. The affirmation as Word produces affirmation as realization or incarnation of the Word, and each of these affirmations corresponds to the denial of its opposite. Thus, in the opinion of the kabbalists, the name of the demon or of evil is composed of the same letters as the name of God or goodness, but spelt backwards. This evil is the last reflection or imperfect mirage of light in shadow. But all which exists, whether of good or evil, in light or darkness, exists and manifests by the tetrad. The affirmation of unity supposes the number four, unless it turns in unity itself as in a vicious circle. So also the triad, as we have already observed, is explained by the duad and resolved by the tetrad, which is the squared unity of even numbers and the quadrangular base of the cube, unity of construction, of solidity and of measure.

The kabbalistic tetragram, Jodheva, expresses God in humanity and humanity in God. The four astronomical cardinal points are, relatively to us, the yea and the nay of light – east and west – and the yea and nay of warmth – south and north. As we have already said, according to the sole dogma of the Kabbalah, that which is in visible nature reveals that which is in the domain of invisible nature, or secondary causes are in strict proportion and analogous to the manifestations of the First Cause. So is this First Cause invariably revealed by the cross – that unity made up of two, that key to the mysteries of India and Egypt, the Tau of the patriarchs, the divine sign of Osiris, the Stauros of the Gnostics, the keystone of the temple, the symbol of Occult Masonry; the cross, central point of the junction of the right angles of two infinite triangles; the cross, which in the French language seems to be the first root and fundamental substantive of the verb to believe and the verb to grow, thus combining the conceptions of science, religion and progress.

The great magical agent manifests by four kinds of phenomena, and has been subjected to the experiments of profane science under four names –

caloric, light, electricity, magnetism. It has received also the names of Tetragram, Inri, Azoth, Ether, Od, Magnetic Fluid, Soul of the Earth, Lucifer, etc. The great magical agent is the fourth emanation of the life-principle, of which the sun is the third form – see the initiates of the school of Alexandria and the dogma of Hermes Trismegistus. In this way the eye of the world, as the ancients called it, is the mirage of the reflection of God, and the soul of the earth is a permanent glance of the sun which the earth conceives and guards by impregnation. The moon concurs in this impregnation of the earth by reflecting a solar image during the night, so that Hermes was right when he said of the great agent: "The sun is its father, the moon its mother." Then he adds: "The wind has borne it in the belly thereof," because the atmosphere is the recipient and, as it were, the crucible of the solar rays, by means of which there forms that living image of the sun which penetrates the whole earth, fructifies it, and determines all that is produced at its surface by its emanations and permanent currents, analogous to those of the sun itself. This solar agent subsists by two contrary forces – one of attraction and one of projection, whence Hermes says that it ascends and descends eternally. The force of attraction is always fixed at the center of bodies, that of projection in their outlines or at their surface. By this dual force all is created and all preserved. Its motion is a rolling up and an unrolling which is successive and indefinite, or, rather, simultaneous and perpetual, by spirals of opposite movements which never meet. It is the same movement as that of the sun, which attracts and repels at once all the planets of its system. To be acquainted with the movement of this terrestrial sun in such a manner as to be able to take advantage of its currents and direct them, is to have accomplished the great work and to be master of the world. Armed with such a force you may make yourself adored; the crowd will believe you are God.

The absolute secret of this direction has been in the possession of certain men, and can yet be discovered. It is the great magical arcanum, depending on an incommunicable axiom and on an instrument which is the grand and unique athanor of the highest grade of Hermetists. The incommunicable axiom is enclosed kabbalistically enclosed in the four letters of the tetragram arranged in the following manner:–

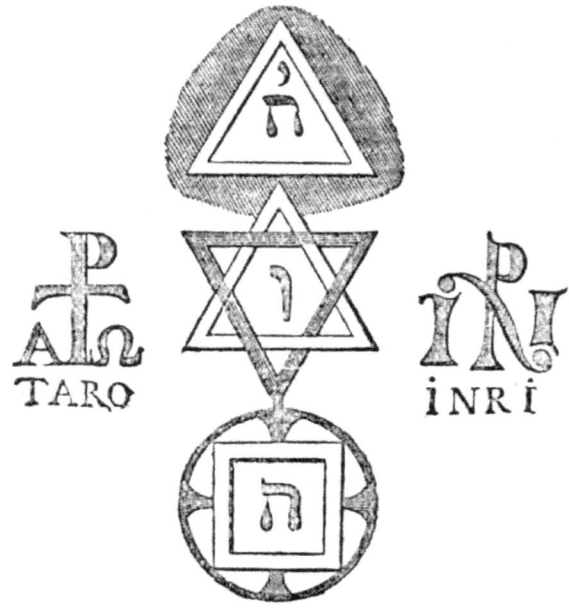

In the letters of the words AZOTH and INRI written kabbalistically; and in the monogram of Christ as embroidered on the labarum, which the Kabbalist Postel interprets by the word ROTA, whence the adepts have formed their Taro or Tarot, by the repetition of the first letter, thus indicating the circle, and suggesting that the word is put backwards. All magical science is comprised in the knowledge of this secret. To know it and have the courage to use it is human omnipotence; to reveal it to a profane person is to lose it; to reveal it even to a disciple is to abdicate in favor of that disciple, who, henceforward, possesses the right of life and death over his master – I am speaking from the magical standpoint – and will certainly slay him for fear of dying himself. But this has nothing in common with deeds qualified as murder in criminal legislation; the practical philosophy which is the basis and point of departure for our laws does not recognize the facts of bewitchment and of occult influences. We touch here upon extraordinary revelations and are prepared for the unbelief and derision of incredulous fanaticism; Voltairean religion has also its fanatics, *pace* the great shades who must now be lurking sullenly in the vaults of the Pantheon, while Catholicism, strong ever in its practices and prestige, chants the office overhead.

The perfect word, that which is adequate to the thought which it expresses, always contains virtually or supposes a tetrad: the idea, with its three necessary and correlated forms, then the image of the thing expressed, with the three terms of the judgement which qualifies it. When I say: "Being exists," I affirm implicitly that the void is nonexistent. A height, a breadth which the height subdivides longitudinally, a depth separated from the height by the intersection of the breadth, such is the natural tetrad composed of two lines at right angles one to another. Nature has also four motions produced by two forces which sustain each other by their tendency in an opposite direction. Now, the law which rules bodies is analogous to that which governs minds, and that which governs minds is the very manifestation of God's secret – that is to say, of the mystery of the creation. Imagine a watch having two parallel springs, with an engagement which makes them work in an opposite direction so that the one in unwinding winds up the other. In this way, the watch will wind up itself, and you will have discovered perpetual motion. The engagement should be at two ends and of extreme accuracy. Is this beyond attainment? We think not. But when it is found out the inventor will understand by analogy all the secrets of nature – *progress in direct proportion to resistance*. The absolute movement of life is thus the perpetual consequence of two contrary tendencies which are never opposed. When one seems to yield to the other, it is a spring which is winding up, and you may expect a reaction, the moment and characteristics of which it is quite possible to foresee and determine. Hence at the period of the greatest Christian fervor was the reign of ANTICHRIST known and predicted. But Antichrist will prepare and determine the second advent and final triumph of the Man-God. This again is a rigorous and kabbalistical conclusion contained in the Gospel premises. Hence the Christian prophecy comprises a fourfold revelation: I. Fall of the old world and triumph of the Gospel under the first advent; 2. Great apostasy and coming of Antichrist; 3. Fall of

Antichrist and recurrence to Christian ideas; 4. Definitive triumph of the Gospel, or Second Advent, designated under the name of the Last Judgement. This fourfold prophecy contains, as will be seen, two affirmations and two negations, the idea of two ruins or universal deaths and of two resurrections; for to every conception which appears upon the social horizon an east and a west, a zenith and a nadir, may be ascribed without fear of error. Thus is the philosophical cross the key of prophecy, and all gates of science may be opened with the pantacle of Ezekiel, the center of which is a star formed by the interlacement of two crosses.

Does not human life present itself also under these four phases or successive transformations – birth, life, death, immortality? And remark here that the immortality of the soul, necessitated as a complement of the tetrad, is kabbalistically proved by analogy, which is the sole dogma of truly universal religion, as it is the key of science and the universal law of nature. As a fact, death can be no more an absolute end than birth is a real beginning. Birth proves the pre-existence of the human being, since nothing is produced from nothing, and death proves immortality, since being can no more cease to be being than nothingness can cease to be nothingness. Being and nothingness are two absolutely irreconcilable ideas, with this difference, that the idea of nothingness, which is altogether negative, issues from the idea itself of being, whence nothingness cannot even be understood as an absolute negation, whilst the notion of being can never be referred to that of nothingness, and still less can it come forth therefrom. To say that the world has been produced out of nothing is to advance a monstrous absurdity. All that is proceeds from what has been, and consequently nothing that is can ever more cease to be. The succession of forms is produced by the alternatives of movement; they are the phenomena of life which replace one another without destroying themselves. All things change; nothing perishes. The sun does not die when it vanishes from the horizon; even the most fluidic forms are immortal, subsisting always in the permanence of their *raison d'être*, which is the combination of light with the aggregated potencies of the molecules of the first substance. Hence they are preserved in the astral fluid, and can be evoked and reproduced according to the will of the sage, as we shall see when treating of second sight and the evocation of memories in necromancy or other magical works. We shall return also to the great magical agent in the fourth chapter of the Ritual, where we shall complete our indications of the characteristics of the great arcanum, and of the means of recovering this tremendous power.

Here let us add a few words on the four magical elements and elementary spirits. The magical elements are: in alchemy, salt, sulphur, mercury and azoth; in Kabbalah, the macroprosopus, the microprosopus and the two mothers; in hieroglyphics, the man, eagle, lion and bull; in old physics, according to vulgar names and notions, air, water, earth and fire. But in magical science we know that water is not ordinary water, fire is not simply fire, etc. These expressions conceal a more recondite meaning. Modern science has decomposed the four elements of the ancients and reduced them to a number of so-called simple bodies. That which is simple, however, is the primitive substance properly so-called; there is therefore only one material element, which manifests always by the tetrad in its forms. We shall therefore preserve the wise distinction of elementary appearances admitted by the ancients, and shall recognize air, fire, earth and water as the four positive and visible elements of magic.

The subtle and the gross, the swift and slow dissolvent, or the instruments of heat and cold, constitute, in occult physics, the two positive and negative principles of the tetrad, and should be thus tabulated:–

Transcendent Magic

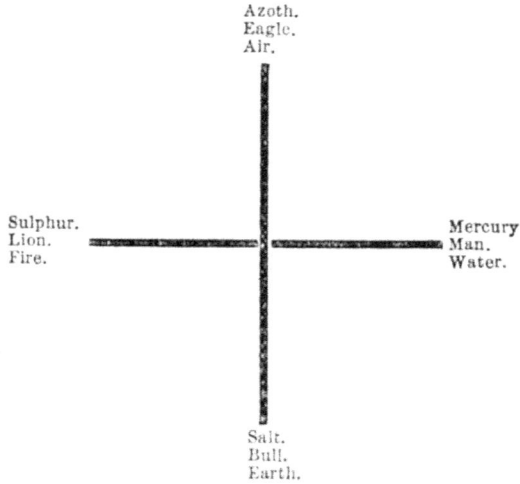

Thus, air and earth represent the male principle; fire and water are referable to the female principle, since the philosophical cross of pantacles, as affirmed already, is a primitive and elementary hieroglyph of the lingam of the gymnosophists. To these four elementary forms correspond the four following philosophical ideas – Spirit, Matter, Motion, Rest. As a fact, all science is comprised in the understanding of these four things, which alchemy has reduced to three – the Absolute, the Fixed and the Volatile – referred by the Kabbalah to the essential idea of God, who is absolute reason, necessity and liberty, a threefold notion expressed in the occult books of the Hebrews. Under the names of Kether, Chokmah and Binah for the Divine World; of Tiphereth, Chesed and Geburah in the moral world, and of Jesod, Hod and Netsah in the physical world, which, together with the moral, is contained in the idea of the Kingdom or Malchuth, we shall explain in the tenth chapter this theogony as rational as it is sublime.

Now, created spirits, being called to emancipation by ordeal, are placed from their birth between these four forces, two positive and two negative, and have it in their power to affirm or deny good, to choose life or death. To discover the fixed point, that is, the moral center of the cross, is the first problem which is given them to resolve; their initial conquest must be that of their own liberty. They begin by being drawn, some to the north, others to the south; some to the right, others to the left; and in so far as they are not free, they cannot have the use of reason, nor can they take flesh otherwise than in animal forms. These unemancipated spirits, slaves of the four elements, are those which the kabbalists call elementary daimons, and they people the elements which correspond to their state of servitude. Sylphs, undines, gnomes and salamanders therefore really exist, some wandering and seeking incarnation, others incarnate and living on this earth. These are vicious and imperfect men. We shall return to this subject in the fifteenth chapter, which treats of enchantments and demons.

That is also an occult tradition by which the ancients were led to admit the existence of four ages in the world, only it was not made known to the vulgar that these ages were successive and were renewed, like the four seasons of the year. Thus, the golden age has passed, and it is yet to come. This, however, belongs to the spirit of prophecy, and we shall speak of it in the ninth chapter, which is concerned with the initiate and the seer. If we now add the idea of unity to the tetrad, we shall have, together and separately, the conceptions of the divine synthesis and analysis, the god of the initiates and that of the profane. Here the doctrine becomes more popular, as it passes from the domain of the abstract; the grand hierophant intervenes.

5 ה E

THE PENTAGRAM

GEBURAH ECCE

Hereunto we have exposed the magical dogma in its more arid and abstruse phases; now enchantments begin; now we can proclaim wonders and reveal most secret things. The pentagram signifies the domination of the mind over the elements, and by its sign are enchained the demons of the air, the spirits of fire, the phantoms of the water, and ghosts of earth. Equipped with this sign, and suitably disposed, you may behold the infinite through the medium of that faculty which is like the soul's eye, and you will be ministered unto by legions of angels and hosts of fiends.

And now, in the first place, let us establish certain principles. There is no invisible world; there are, however, many degrees of perfection in organs. The body is the course and, as it were, the perishable cortex of the soul. The soul can perceive of itself, and independently of the mediation of the physical organs, by means of its sensibility and its diaphane, the things, both spiritual and corporeal, which are existent in the universe. Spiritual and corporal are simply terms which express the degrees of tenuity or density in substance. What is called the imagination within us is only the soul's inherent faculty of assimilating the images and reflections contained in the living light, which is the great magnetic agent. These images and reflections are revelations when science intervenes to reveal us their body or light. The man of genius differs from the dreamer and the fool in this only, that his creations are analogous to truth, while those of the fool and the dreamer are lost reflections and bewrayed images. Hence, for the wise man, to imagine is to see, as, for the magician, to speak is to create. Therefore, by means of the imagination, demons and spirits can be beheld really and in truth; but the imagination of the adept is diaphanous, whilst that of the crowd is opaque; the light of truth traverses the one as ordinary light passes through a transparent casement, and is refracted by the other as when ordinary light falls upon a vitreous block full of scoria and foreign matter. That which most contributes to the errors of the vulgar is the reflection of depraved imaginations one in the other. But the seer, by a positive science, knows that what he imagines is true, and the event invariably confirms his vision. We shall state in the Ritual after what manner this lucidity can be acquired.

It is by means of this light that static visionaries place themselves in communication with all worlds, as so frequently occurred to Swedenborg, who, notwithstanding, was imperfectly lucid, seeing that he did not distinguish reflections from rays, and often intermingled chimerical fancies with his most admirable dreams. We say dreams, because dream is the consequence of a natural and periodical ecstasy, which we term sleep; to be in ecstasy is to sleep; magnetic somnambulism is a production and direction of sleep. The errors which occur therein are occasioned by reflections from the diaphane of waking persons, and, above all, of the magnetizer. Dream is vision produced by the refraction of a ray of truth. The chimerical fantasy is hallucination occasioned by a reflection. The temptation of St. Anthony, with its nightmares and its monsters, represents the confusion of reflections with direct rays. So long as the soul struggles it is reasonable; when it yields to this specie, of invading intoxication it becomes mad. To disentangle the direct ray, and separate it from the reflection – such is the work of the initiate. Here let us state distinctly that this work is through all times accomplished in the world by some of the flower of mankind, that there is hence a permanent revelation by intuition, and that there is no insuperable barrier which separates souls, because there are no sudden interruptions, and no high walls in nature by which minds can be divided from one another. All is transition and blending, and, assuming the perfectibility, if not infinite, at least indefinite, of human faculties, it will be seen that every person can attain to see all, and therefore to know all. There is no void in nature; all is peopled. There is no true death in nature; all is alive. "Seest thou that star?" asked Napoleon of Cardinal Fesch. "No, Sire." "I see it," said the Emperor, and he most certainly did. When great men are accused of having been superstitious, it is because they behold what remains unseen by the crowd. Men of genius differ from simple seers by their faculty of sensibly communicating to other men what they themselves perceive, and of making themselves believed by the force of enthusiasm and sympathy. Such persons are the medium of the Divine Word.

Let us now state the manner in which visions operate. All forms correspond to ideas, and there is no idea which has not its proper and peculiar form. The primordial light, which is the vehicle of all ideas, is the mother of all forms, and transmits them from emanation to emanation, merely diminished or modified according to the density of the media.

Secondary forms are reflections which return to the font of the emanated light. The forms of objects, being a modification of light, remain in the light where the reflection consigns them. Hence the Astral Light, or terrestrial fluid, which we call the great magnetic agent, is saturated with all kinds of images or reflections. Now, our soul can evoke these, and refer them to its diaphane, as the kabbalists term it. Such images are always present to us, and are only effaced by the more powerful impressions of reality during waking hours, or by preoccupation of the mind, which makes our imagination inattentive to the fluidic panorama of the astral light. When we sleep, this spectacle presents itself spontaneously before us, and in this way, dreams are produced – dreams vague and incoherent if some governing will do not remain active during the sleep, giving, even unconsciously to our intelligence, a direction to the dream, which then transforms into vision. Animal magnetism is nothing else but an artificial sleep produced by the voluntary or enforced union of two wills, one of which is awake while the other slumbers – that is, one of which directs the other in the choice of reflections for the transformation of dreams into visions, and the attainment of truth by means of images. Thus, somnambulists do not actually travel to the place where they are sent by the magnetizer; they evoke its images in the astral light and can behold nothing which does not exist in that light. The astral light has a direct action on the nerves, which are its conductors in the animal economy, transmitting it to the brain, whence also, in the state of somnambulism, it is possible to see by means of the nerves, without being dependent on radiant light, the astral fluid being a latent light, in the same way that physics recognize the existence of a latent caloric.

Magnetism between two persons is certainly a wonderful discovery, but the magnetizing of a person by himself, accomplishing his own lucidity and directing himself at will, is the perfection of magical art. The secret of this great work does not rest for discovery; it has been known and practiced by a great number of initiates, above all by the celebrated Apollonius of Tyana, who has left a theory concerning it, as we shall see in the Ritual. The secret of magnetic lucidity, and the direction of the phenomena of magnetism depend on two things – the agreement of minds and the complete union of wills, in a direction which is possible and determined by science. This is for the operation of magnetism between two or more persons. Solitary magnetism requires preparations of which we have spoken in our initial chapter, when enumerating and establishing in all their difficulty the essential qualities of a veritable adept. In the following chapters we shall elucidate further this important and fundamental point.

The empire of will over the astral light, which is the physical soul of the four elements, is represented in magic, by the pentagram, which we have set at the head of this chapter. The elementary spirits are subservient to this sign when employed with understanding, and, by placing it in the circle or on the table of evocations, they can be rendered tractable, which is magically called to imprison them. Let us briefly explain this marvel. All created beings communicate with one another by signs, and all adhere to a certain number of truths expressed by determinate forms. The perfection of forms increases in proportion to the detachment of spirits, and those that are not over weighted by the chains of matter, recognize by intuition out of hand whether a sign is the expression of a real power or of a precipitate will. The intelligence of the wise man therefore gives value to his pantacle, as science gives weight to his will, and spirits comprehend this power immediately. Thus, by means of the Pentagram, spirits can be forced to appear in vision, whether in the waking or sleeping state, *by themselves leading before our diaphane their reflection, which exists in the astral light, if they have lived, or a reflection analogous to their spiritual logos if they have not lived on earth.* This explains all visions, and accounts for the dead invariably appearing to seers, either such as they were upon earth, or such as they are in the grave, never as they subsist in a condition which escapes the perceptions of our actual organism.

Pregnant women are influenced more than others by the astral light, which concurs in the formation of the child, and perpetually offers them reminiscences of the forms which abound therein. This explains how it is that women of the highest virtue deceive the malignity of observers by equivocal resemblances. On the fruit of their marriage they impress frequently an image which has struck them in dream, and it is thus that the same physiognomies are perpetuated from generation to generation. The Kabbalistic usage of the pentagram can therefore determine the appearance of unborn children, and an initiated woman might endow her son with the characteristics of Nero or Achilles, with those of Louis XIV. or Napoleon. We shall indicate the method in our Ritual.

The Pentagram is called in Kabbalah the sign of the microcosm, that sign so exalted by Goethe in the beautiful monologue of Faust: "Ah, how do all my senses leap at this sight! I feel the young and sacred pleasure of life bubbling in my nerves and veins. Was it a God who traced this sign which stills the vertigo of my soul, fills my poor heart with joy, and, in a mysterious rapture, unveils the forces of Nature around me. Am I myself a God! All is so clear to me; I behold in these simple lines the revelation of active nature to my soul. I realize for the first time the truth of the wise man's words: The world of spirits is not closed! Thy sense is obtuse, thy heart is dead! Arise! Bathe, O adept of science, thy breast, still enveloped by an earthly veil, in the splendors of the dawning day!" (Faust, Part I. sc. I.)

On the 24th of July in the year 1854, the author of this book, Éliphas Lévi, made experiments of evocation with the pentagram, after due preparation by all the Ceremonies which are indicated in the thirteenth chapter of the *Ritual*. The success of this experiment, details of which, as regards its principles, will be found in the corresponding chapter of this our doctrinal part, establishes a new pathological fact, which men of true science will admit without difficulty. The repeated experience, in all three times, gave results truly extraordinary, but positive and unmixed with hallucination. We invite sceptics to make a conscientious and intelligent attempt before

Transcendent Magic

shrugging their shoulders and smiling. The figure of the pentagram, perfected in accordance with science, and used by the author in his experiment, is that which is found at the head of this chapter, and it is more perfect than any in the keys of Solomon or in the magical calendars of Tycho Brahe and Duchentau. We must, however, remark that the use of the pentagram is most dangerous for operators who are not in possession of its complete and perfect understanding. The direction of the points of the star is in no sense arbitrary, and may change the entire character of the operation, as we shall explain in the Ritual.

Paracelsus, that innovator in magic, who surpassed all other initiates in his unaided practical success, affirms that every magical figure and every kabbalistic sign of the pantacles which compel spirits, may be reduced to two, which are the synthesis of all the others; these are the sign of the Macrocosm or the seal of Solomon, the form of which we have already given, and now reproduce here, and that of the Microcosm, more potent even than the first – that is to say, the pentagram, of which he provides a most minute description in his occult philosophy. If it be asked how a sign can exercise so much power over spirits, we inquire in return why the whole Christian world bows down before the sign of the cross? The sign is nothing by itself, and has no force apart from the doctrine of which it is the summary and the logos. Now, a sign which sums, by their expression, all the occult forces of nature, a sign which has ever exhibited to elementary spirits and others a power greater than their own, naturally fills them with respect and fear, and enforces their obedience by the empire of science and of will over ignorance and weakness. By the pentagram also is measured the exact proportions of the great and unique athanor necessary to the confection of the philosophical stone and the accomplishment of the great work. The most perfect alembic in which the quintessence can be elaborated is conformable to this figure, and the quintessence itself is represented by the sign of the pentagram.

6 ו F

MAGICAL EQUILIBRIUM

TIPHERETH UNCUS

SUPREME intelligence is necessarily reasonable. God, in philosophy, may be only a hypothesis, but he is a hypothesis imposed by good sense on human reason. To personify the Absolute Reason is to determine the divine ideal. Necessity, liberty and reason – these are the great and supreme triangle of the Kabbalists, who name reason Kether, necessity Chokmah, and liberty Binah, in their first or divine triad. Fatality, will and power, such is the magical triad, which corresponds in things human to the divine triad. Fatality is the inevitable sequence of effects and causes in a determined order. Will is the directing faculty of intelligent forces for the conciliation of the liberty of persons with the necessity of things. Power is the wise application of will which enlists fatality itself in the accomplishment of the desires of the sage. When Moses smote the rock, he did not create the spring of water, he revealed it to the people, because occult science had made it known to himself by means of the divining rod. It is in like manner with all miracles of magic; a law exists, which is ignored by the vulgar and made use of by the initiate. Occult laws are often opposed diametrically to common ideas. For example, the crowd believes in the sympathy of things which are alike and in the hostility of things contrary, but it is the opposite which is the true law. It used to be affirmed that nature detests the void, but it should be said that nature desires it, were the void not, in physics, the most irrational of fictions. In all things the vulgar mind habitually takes shadow for reality, turns its back upon light, and is reflected in the obscurity which it projects itself. The forces of Nature are at the disposal of one who knows how to resist them. Are you master sufficiently of yourself to be never intoxicated? Then will you direct the terrible and fatal power of intoxication. If you would make others drunk, possess them with the desire of drink, but do not partake of it yourself. That man will dispose of the love of others who is master of his own. If you would possess, do not give. The world is magnetized by the light of the sun, and we are magnetized by the astral light of the world. That which operates in the body of the planet repeats itself in us. Within us there are three analogical and hierarchic worlds, as in all nature.

Man is the microcosm or little world, and, according to the doctrine of analogies, whatsoever is in the great world is reproduced in the small. Hence we have three centers of fluidic attraction and projection – the brain, the heart or epigastric region, and the genital organ. Each of these instruments is double – in other words, we find the suggestion of the triad therein. Each attracts on one side and repels on another. It is by means of these apparatuses that we place ourselves in communication with the universal fluid transmitted into us by the nervous system. These three centers are, moreover, the seat of a threefold magnetic operation, as we shall explain elsewhere. When the magus has attained lucidity, whether through the mediation of a pythoness, or by his own development, he communicates and directs at will the magnetic vibrations in the whole mass of the astral light, the currents of which he divines by means of the magic rod, which is a perfected divining rod. By the aid of these vibrations he influences the nervous system of persons surrendered to his action, accelerates or suspends the currents of life, soothes or tortures, heals or hurts; in fine, slays or brings to life. . . . Here, however, we pause in presence of the smile of incredulity. Let us permit it to enjoy the cheap triumph of denying what it does not know.

We shall demonstrate later on that death is always preceded by a lethargic sleep, and only takes place gradually; that resurrection is possible in certain cases; that lethargy is a real, but uncompleted, death; and that the final paroxysm is in many cases subsequent to inhumation. This, however, is not the subject of the present chapter. We now affirm that a lucid will can act upon the mass of the astral light, and, in concurrence with other wills, which it absorbs and draws along, can determine great and irresistible currents. We say also that the astral light condenses or rarefies in proportion as currents accumulate, more or less, at certain centers. When it is deficient in the energy required for the support of life, diseases accompanied by sudden decomposition follow, of the kind which baffle physicians. There is no other cause, by example, in the case of cholera-morbus, and the swarms of animalculæ observed or supposed by some specialists may be the effect rather than the cause. Cholera should therefore be treated by insufflation, did not the operator thereby run the chance of an exchange with the patient, which would be very formidable for himself. Every intelligent effort of will is a projection of the human fluid or light, and here it is needful to distinguish the human from the astral light, and animal from universal magnetism. In making use of the word fluid, we employ an accepted expression, and would make ourselves understood in this manner, but we are far from deciding that the latent light is a fluid. Everything prompts us, on the contrary, to prefer the system of vibrations in the explanation of this phenomenal subject. However it may be, the light in question, being the instrument of life, cleaves naturally to all living centers, attaches itself to the nucleus of planets, even as to the heart of man – and by the heart we understand magically the great sympathetic – identifying itself with the individual life of the being which it animates, and it is by this quality of sympathetic assimilation that it distributes itself without confusion. Hence it is terrestrial in its affinity with the sphere of the earth, and human exclusively in its affinity with men.

It is for this reason that electricity, caloric, light, and magnetism, produced by ordinary physical means, not only do not originate, but rather tend to neutralize the effects of animal magnetism. The astral light, subordinated to a blind mechanism, and proceeding from arbitrary automatic centers, is a dead light and works mathematically, following given impulsions or fatal laws; the human light is fatal only to the ignorant in chance experiments; in the seer it is subjected to intelligence, submitted to imagination, and dependent on will. This light, continually projected by the will, constitutes the personal atmospheres of Swedenborg. The body absorbs what environs it, and radiates perpetually by projecting its influences and invisible molecules; it is the same with the spirit, so that this phenomenon, by some mystics termed *respiration*, has really the influence, both physical and moral, which is assigned to it. It is undoubtedly contagious to breathe the same air as diseased persons, and to be within the circle of attraction and expansion which surrounds the wicked.

When the magnetic atmosphere of two persons is so equilibrated that the attractive faculty of one draws the expansive faculty of the other, a tendency is produced which is termed sympathy; then imagination, calling up to it all the rays or reflections analogous to that which it experiences, makes a poem of the desires which captivate will, and if the persons differ in sex, it occasions in them, or more commonly in the weaker of the two, a complete intoxication of the astral light, which is termed passion *par excellence,* or love. Love is one of the great instruments of magical power, but it is categorically forbidden to the magus, at least as an intoxication or passion. Woe to the Samson of the Kabbalah if he permits himself to be put asleep by Delilah! The Hercules of science, who exchanges his royal scepter for the distaff of Omphalô, will soon experience the vengeance of Dejanira, and nothing will be left for him but the pyre of Mount Œtá, in order to escape the devouring folds of the coat of Nessus. Sexual love is ever an illusion, for it is the result of an imaginary mirage. The astral light is the universal seducer, typified by the serpent of Genesis. This subtle agent, ever active, ever abounding in sap, ever fruitful in alluring dreams and sensuous images; this force, which of itself is blind and subordinated to every will, whether for good or evil; this ever-renewing *circulus* of unbridled life, which produces vertigo in the imprudent; this corporeal spirit; this fiery body; this impalpable omnipresent ether; this monstrous seduction of nature – how shall we define it comprehensively and how characterize its action? To some extent indifferent in itself, it lends itself to good as to evil; it transmits light and propagates darkness; it may be called equally Lucifer and Lucifuge; it is a serpent but it is also an aureole; it is a fire, but it may belong equally to the torments of infernus or the sacrifice of incense offered up to heaven. To dispose of it, we must, like the predestined woman, set our foot upon its head.

In the elementary world water corresponds to the kabbalistic woman and fire to the serpent. To subdue the serpent, that is, to govern the circle of the astral light, we must place ourselves outside its currents, that is, we must isolate ourselves. For this reason Apollonius of Tyana wrapped himself completely in a mantle of fine wool, setting his feet thereon and drawing it over his head. Then he bent his back in semicircular fashion, and closed his eyes, after performing certain rites, probably magnetic passes and sacramental words, designed to fix imagination and determine the action of the will. The woolen mantle is of great use in magic and was the common conveyance of sorcerers on their way to the Sabbath, which proves that the sorcerers did not really go to the Sabbath, but the Sabbath came to the sorcerers, when isolated in their mantle, and brought before their *translucid* images analogous to their magical preoccupations, combined with reflections of all kindred acts previously accomplished in the world.

This torrent of universal life is also represented in religious doctrines by the expiatory fire of hell. It is the instrument of initiation, the monster to be overcome, the enemy to subdue; it is this which brings to our evocations and to the conjurations of goëtic magic such swarms of larvæ and phantoms; therein are preserved all the forms by which their fantastic and fortuitous assemblage people our nightmares with such abominable deformities. To allow ourselves to be sucked down by this whirling stream is to fall into the abysses of madness, more frightful than those of death; to expel the darkness of this chaos and force it to give perfect forms to our thoughts – this is, to be a man of genius, it is to create, it is to be victorious over hell! The astral light directs the instincts of animals and offers battle to the intelligence of man, which it strives to pervert by the enticements of its reflections, and the illusion of its images, a fatal and inevitable operation, directed and made still more calamitous by elementary spirits and suffering souls, whose restless wills seek out sympathies in our weakness, and tempt us not so much to destroy us as to win friends for themselves.

That book of consciences which, according to Christian doctrine, shall be opened at the last day, is no other than the Astral Light, which preserves the impress of every logos, that is to say, of all actions and all forms. Our acts modify our magnetic respiration in such a way that a seer, meeting any person for the first time, can tell whether he is innocent or criminal, and what are his virtues or his crimes. This faculty, which belongs to divination, was called by the Christian mystics of the early Church the discernment of spirits.

Those who abdicate the empire of reason and delight to let their wills wander in pursuit of the reflections in the astral light, are subject to alternations of mania and melancholy which have originated all the marvels of demoniacal possession, though it is true, at the same time, that by means of these reflections impure spirits can act upon similar souls, make use of them as docile instruments, and even habitually torment their organism, wherein they enter and reside by *obsession*, or *embryonically*. These kabbalistic terms are explained in the Hebrew book of the Revolution of Souls, of which our thirteenth chapter will contain a succinct analysis. It is therefore extremely dangerous to make sport of the mysteries of magic; it is, above all, excessively rash to practice its rites from curiosity, by experiment, and as if to exploit higher forces. The inquisitive who, without being adepts, busy themselves with evocations or occult magnetism, are like children playing with fire in the neighborhood of a cask of gunpowder; sooner or later they will fall victims to some terrible explosion.

To be isolated from the astral light it is not enough to envelop one's self in a woolen fabric; we must also, and above all, impose absolute tranquility on mind and heart, we must have quitted the world of passions and be assured of perseverance in the spontaneous operations of an inflexible will. We must reiterate frequently the acts of this will, for, as we shall see in the introduction to the Ritual, the will only assures itself by acts, as the power and perpetuity of religions depend on their rites and ceremonies.

There are intoxicating substances, which, by increasing nervous sensibility, exalt the power and consequently the allurements of astral representations; by the same means, but pursuing a contrary course, spirits may be terrified and disturbed. These substances, of themselves magnetic, and further magnetized by operators, are what people term philtres and enchanted potions. But we shall not enter here upon this dangerous application of magic, which Cornelius Agrippa himself terms venomous magic. It is true that there are no longer pyres for sorcerers, but always, and more than ever, are there penalties dealt out to malefactors. Let us confine ourselves therefore to stating, as the occasion offers, the reality of this power.

To direct the astral light, we must understand also its double vibration, as well as the balance of forces termed magical equilibrium and expressed in the Kabbalah by the senary[12]. Considered in its first cause, this equilibrium is the will of God; it is liberty in man, and mathematical equilibrium in matter. Equilibrium produces stability and duration.

---

[12] Spelled exactly as in the printed copy. -pnw

Liberty generates the immortality of man, and the will of God gives effect to the laws of eternal reason. Equilibrium in ideas is reason and in forces power. Equilibrium is exact; fulfil its law, and it is there; violate it, however slightly, and it is destroyed. For this reason, nothing is useless or lost. Every utterance and every movement are for or against truth, which is composed of *for* and *against* conciliated, or at least equilibrated. We shall state in the introduction to the Ritual how magical equilibrium should be produced, and why it is necessary to the success of all operations.

Omnipotence is the most absolute liberty; now, absolute liberty cannot exist apart from perfect equilibrium. Magical equilibrium is hence one of the first conditions of success in the operations of science, and must be sought even in occult chemistry, in learning to combine contraries without neutralizing them by one another. Magical equilibrium explains the great and primeval mystery of the existence and relative necessity of evil. This relative necessity gives, in black magic, the measure of the power of demons or impure spirits, to whom virtues practiced upon earth are a source of increased rage and apparently of increased power. At the epochs when saints and angels work miracles openly, sorcerers and fiends in their turn operate marvels and prodigies. Rivalry often creates success; we lean upon that which resists.

7 ו G

THE FIERY SWORD

NETSAH GLADIUS

The septenary is the sacred number in all theogonies and in all symbols, because it is composed of the triad and the tetrad. The number seven represents magical power in all its fullness; it is the mind reinforced by all elementary potencies; it is the soul served by nature; it is the *sanctum regnum* mentioned in the keys of Solomon, and represented in the Tarot by a crowned warrior, who bears a triangle on his cuirass, and is posed upon a cube, to which two sphinxes are harnessed, straining in opposite directions, while their heads are turned the same way. This warrior is armed with a fiery sword, and holds in his other hand, a scepter surmounted by a triangle and a sphere. The cube is the philosophical stone; the sphinxes are the two forces of the great agent, corresponding to Jakin and Bohas, the two pillars of the Temple; the cuirass is the knowledge of divine things, which renders the wise man invulnerable to human assaults; the scepter is the magic rod; the fiery sword is the symbol of victory over the deadly sins, seven in number, like the virtues, the conceptions of both being typified by the ancients under the figures of the seven planets then known. Thus, faith – that aspiration towards the infinite, that noble self-reliance sustained by confidence in all virtues – that faith, which in weak natures, may degenerate into pride, was represented by the Sun; hope, the enemy of avarice, by the Moon; charity, in opposition to luxury, by Venus, the bright star of morning and evening; strength, superior to wrath, by Mars; prudence, hostile to idleness, by Mercury; temperance, opposed to gluttony, by Saturn, who was given a stone instead of his children to devour; finally, justice, in opposition to envy, by Jupiter, the conqueror of the Titans. Such are the symbols borrowed by astronomy from the Hellenic cultus. In the Kabbalah of the Hebrews, the Sun represents the angel of light; the Moon, the angel of aspirations and dreams; Mars, the destroying angel; Mercury, the angel of progress; Jupiter, the angel of power; Saturn, the angel of the wilderness.[13] They were named also Michaël, Gabriel, Samaël, Anaël, Raphaël, Zachariel, and Orifiel.

These governing potencies of souls shared human life in periods, which astrologers measured by the revolutions of the corresponding planets. But kabbalistic astrology must not be confounded with judicial astrology. We will explain this distinction. Infancy is dedicated to the Sun, childhood to the Moon, youth to Mars and Venus, manhood to Mercury, ripe age to Jupiter, and old age to Saturn. Now, humanity in general subsists under laws of development analogous to those of individual life. On this groundwork Trithemius establishes his prophetic key of the seven spirits, to which we shall subsequently refer; by means thereof, observing the analogical proportions of successive events, it is possible to predict important future occurrences with certitude, and to fix beforehand, from age to age, the destinies of nations and the world. St. John, depositary of the secret doctrine of Christ, has introduced it into the kabbalistic book of the Apocalypse, which he represents sealed with seven seals. We there find the seven genii of ancient mythologies, with the Cups and Swords of the Tarot. The doctrine concealed under these emblems is pure Kabbalah, already lost by the Pharisees at the time of Christ's advent. The scenes which succeed one another in this wonderful prophetic epic are so many pantacles, the keys of which are the ternary, the quaternary, the septenary,

---

[13] Another version of this text includes, "Venus, the angel of loves" -pnw

and the duodenary. Its hieroglyphic figures are analogous to those of the Book of Hermes or the Genesis of Enoch, to make use of a tentative title which expresses merely the personal opinion of the erudite William Postel.

The cherub, or symbolic bull, which Moses placed at the gate of the Edenic world, bearing a fiery sword, is a sphinx, having a bull's body and a human head; it is the antique Assyrian sphinx, and the combat and victory of Mithras were its hieroglyphic analysis. Now, this armed sphinx represents the law of mystery which watches at the door of initiation to warn away the profane. Voltaire, who knew nothing of all this, was highly diverted at the notion of a bull brandishing a sword. What would he have said had he visited the ruins of Memphis and Thebes, and what would the echo of past ages which slumbers in the tombs of Rameses have replied to those light sarcasms so much relished in France? The Mosaic cherub represents also the great magical mystery, of which the elements are expressed by the septenary, without, however, giving the final word. This *verbum inenarrabile* of the sages of the Alexandrian school, this word which Hebrew Kabalists write יהוה and interpret by אראריתא, thus expressing the triplicity of the secondary principle, the dualism of the means, and the equal unity of the first and final principle, then further the alliance between the triad and the tetrad in a word composed of four letters, which form seven by means of a triple and double repetition – this word is pronounced Ararita.

The virtue of the septenary is absolute in magic, for this number is decisive in all things; hence all religions have consecrated it in their rites. The seventh year was a jubilee among the Jews; the seventh day is set apart for rest and prayer; there are seven sacraments, etc. The seven colors of the prism and the seven musical notes, correspond also to the seven planets of the ancients, that is, to the seven chords of the human lyre. The spiritual heaven has never changed, and astrology has been more invariable than astronomy. The seven planets are, in fact, the hieroglyphic symbols of the key of our affections. To compose talismans of the Sun, Moon or Saturn, is to attach the will magnetically to signs corresponding to the chief powers of the soul; to consecrate something to Mercury or Venus is to magnetize that object according to a direct intention, whether pleasure, science or profit be the end in view. The analogous metals, animals, plants and perfumes are auxiliaries to this end. The seven magical animals are: (a) Among birds, corresponding to the divine world, the swan, the owl, the vulture, the dove, the stork, the eagle and the pewit; (b) among fish, corresponding to the spiritual or scientific world, the seal, the catfish, the pike, the mullet, the chub, the dolphin, the sepia or cuttlefish; (c) among quadrupeds, corresponding to the natural world, the lion, the cat, the wolf, the he-goat, the monkey, the stag, and the mole. The blood, fat, liver and gall of these animals serve in enchantments; their brain combines with the perfumes of the planets, and it is recognized by ancient practice that they possess magnetic virtues corresponding to the seven planetary influences.

The talismans of the seven spirits are engraved either on precious stones, such as the carbuncle, crystal, diamond, emerald, agate, sapphire and onyx, or upon metals, such as gold, silver, iron, copper, fixed mercury, pewter and lead. The kabbalistic signs of the seven spirits are: for the Sun, a serpent with the head of a lion; for the Moon, a globe divided by two crescents; for Mars, a dragon biting the hilt of a sword; for Venus, a lingam; for Mercury, the Hermetic caduceus and the cynocephalus; for Jupiter, the blazing pentagram in the talons or beak of an eagle; for Saturn, a lame and aged man, or a serpent curled about the sun-stone. All these symbols are found on graven stones of the ancients, and especially on those talismans of the Gnostic epochs which are known by the name of Abraxas. In the collection of the talismans of Paracelsus, Jupiter is represented by a priest in ecclesiastical garb, while in the Tarot he appears as a grand hierophant crowned with a triple tiara, holding a three-fold cross in his hands, forming the magical triangle, and representing at once the scepter and key of the three worlds.

By combining all that has been said about the unity of the triad and tetrad, we shall find all that remains for us to say concerning the septenary, that grand and complete magical unity composed of four and three.[14]

---

[14] With reference to the plants and colors of the septenary employed in magnetic experiences, see the erudite work of M. Ragon on *La Maçonnerie Occulte*.

8 ה H

REALIZATION

HOD VIVENS

CAUSES manifest by effects, and effects are proportioned to causes. The divine word, the one word, the tetragram, has affirmed itself by tetradic creation. Human fecundity proves divine fecundity; the jod of the divine name is the eternal virility of the First Principle. Man understands that he was made in the image of God when he attains comprehension of God by increasing to infinity the idea which he forms of himself. When realizing God as the infinite man, man says unto himself: I am the finite God. Magic differs from mysticism because it judges nothing *à priori* until after it has established *à posteriori* the base itself of its judgements, that is to say, after having understood the cause by the effects contained in the very energy of the cause, by means of the universal law of analogy. Hence in the occult sciences all is real, and theories are established only on the foundations of experience. Realities alone constitute the proportions of the ideal, and the magus admits nothing as certain in the domain of ideas save that which is demonstrated by realization. In other words, what is true in the cause manifests in the effect. What is not realized does not exist. The realization of speech is the logos properly so called. A thought realizes itself in becoming speech; it is realizes itself also by signs, sounds and representations of signs: this is the first degree of realization. Then it is imprinted on the astral light by means of the signs of writing or speech; it influences other minds by reflection upon them; it is refracted by crossing the diaphane of other men; it assumes new forms and proportions; it is then translated into acts and modifies the world: this is the last degree of realization. Men who are born into a world modified by an idea bear away with them the impression thereof, and it is thus that the word is made flesh. The impression of the disobedience of Adam, preserved in the astral light, could be effaced only by the stronger impression of the obedience of the Savior, and thus the original sin and redemption of the world can be explained in a natural and magical sense. The astral light, or soul of the world, was the instrument of Adam's omnipotence; it became afterwards the instrument of his punishment, being corrupted and troubled by his sin, which intermingled an impure reflection with those primitive images which composed the book of universal science for his still virgin imagination.

The astral light, depicted in ancient symbols by the serpent devouring its tail, represents alternately malice and prudence, time and eternity, tempter and Redeemer; for this light, being the vehicle of life, is an auxiliary alike of good and evil, and may be taken for the fiery form of Satan as for the body of the Holy Ghost. It is the instrument of warfare in angelic battles, and indifferently feeds the flames of hell and the lightnings of St. Michael. It may be compared to a horse having a nature analogous to the chameleon, ever reflecting the armor of his rider. The astral light is the realization or form of intellectual light, as the latter is the realization or form of the divine light.

The great initiator of Christianity, divining that the astral light was overcharged with the impure reflections of Roman debauchery, sought to separate his disciples from the ambient sphere of reflections, and to make them attentive only to the interior light, so that, through the medium of a common faith and enthusiasm, they might communicate together by new magnetic chains, which he termed grace, and thus overcome the dissolute currents, to which he gave the names of the devil and Satan, signifying their putrefaction. To oppose current to current is to renew the power of fluidic life. The revealers have, therefore, done scarcely more than divine, by the accuracy of their calculations, the appropriate moment for moral reactions. The law of realization produces what we call magnetic breathing; places and objects become impregnated therewith, and this communicates to them an influence in conformity with our dominant desires, with those above all, which are confirmed and realized by acts. As a fact, the universal agent, or latent astral light, ever seeks equilibrium; it fills the void and sucks up the plenitude, which makes vice contagious, like certain physical maladies, and works powerfully in the proselytism of virtue. Hence it is that cohabitation with antipathetic beings is a torment; hence it is that relics, whether of saints or of great criminals, produce the extraordinary results of sudden conversion and perversion; hence it is that sexual love is often awakened by a breath or a touch, and this, not only by means of the contact of the person himself, but of objects which he has unconsciously touched or magnetized.

There is an outbreathing and inbreathing of the soul, exactly like that of the body. It breathes in the felicity which it believes, and it breathes forth ideas which result from its inner sensations. Diseased souls have an evil breath, and vitiate their moral atmosphere – that is, they combine impure reflections with the astral light which permeates them, and establish unwholesome currents therein. We are often invaded, to our astonishment, in society by evil thoughts which would have seemed impossible, and are not aware that they are due to some morbid proximity. This secret is of high importance, for it leads to the opening of consciences, one of the most incontestable and terrible powers of magical art. Magnetic respiration produces about the soul a radiation of which it is the center, and surrounds it with the reflection of its own works, creating for it a heaven or a hell. There are no isolated acts, and it is impossible that there should be secret acts; whatsoever we truly will – that is, everything which we confirm by our acts – remains registered in the astral light, where our reflections are preserved. These reflections continually influence our thought by the mediation of the diaphane, and it is in this sense that we become and remain the children of our works.

The astral light, transformed at the moment of conception into human light, is the soul's first envelope, and, in combination with extremely subtle fluids, it forms the ethereal body or sidereal phantom, of which Paracelsus discourses in his philosophy of intuition – *philosophia sagax*. This sidereal body, setting itself free at death, attracts, and for a long time preserves, through the sympathy of things homogeneous, the reflections of the past life; if drawn along a special current by a will which is powerfully sympathetic, it manifests naturally, for there is nothing more natural than prodigies. It is thus apparitions are produced. But we shall develop this point more fully in a chapter devoted to Necromancy. The fluidic body, subject, like the mass of the astral light, to two contrary movements, attracting on the left and repelling on the right, or reciprocally, between the two sexes, begets various impulses within us, and contributes to solicitudes of conscience; it is frequently influenced by reflections of other minds, and thus are produced, on the one hand, temptations, and, on the other, profound and unexpected graces. This is also the explanation of the traditional doctrine of two angels who strengthen and tempt us. The two forces of the astral light may be represented by a balance wherein are weighed our good intentions for the triumph of justice and the emancipation of our liberty.

The astral body is not always of the same sex as the terrestrial, that is, the proportions of the two forces, varying from right to left, seem frequently to gainsay the visible organization, producing seeming vagaries of human passion and explaining, without in any sense morally justifying, the amorous peculiarities of Anacreon or Sappho. A skillful magnetizer should take all these subtle distinctions into account, and we shall provide in our Ritual the rules for their recognition.

There are two kinds of realization, the true and the fantastic. The first is the exclusive secret of magicians, the other belongs to enchanters and sorcerers. Mythologies are fantastic realizations of religious dogma; superstitions are the sorcery of mistaken piety; but even mythologies and superstitions are more efficacious with human will than a purely speculative philosophy apart from any practice. Hence St. Paul opposes the conquests of the folly of the Cross to the inertness of human wisdom. Religion *realizes* philosophy by *adapting* it to the weaknesses of the vulgar; such is for Kabbalists the secret reason and occult explanation of the doctrines of incarnation and redemption.

Thoughts untranslated into speech are thoughts lost for humanity; words unconfirmed by acts are idle words, and the idle word is not far removed from falsehood. Thought formulated by speech and confirmed by acts constitutes a good work or a crime. Hence, whether in vice or virtue, there is no speech for which we are not responsible; above all, there are no indifferent acts. Curses and blessings invariably produce their consequence invariably, and every action, whatsoever its nature, whether inspired by love or hate, has effects analogous to its motive, its extent, and its direction. When that emperor whose images had been mutilated, raising his hand to his face, exclaimed, "I do not feel that I am injured," he was mistaken in his valuation, and thereby detracted from the merit of his clemency. What man of honor could behold undisturbed an insult offered to his portrait? And did such insults, inflicted even unknown to ourselves, react on us by a fatal influence, were the effects of bewitchment actual, as indeed an adept cannot doubt, how much more imprudent and ill-advised would seem this utterance of the good emperor!

There are persons whom we can never offend with impunity, and if the injury we have done them is mortal, we forthwith begin to die. There are those also whom we never meet in vain, whose mere glance alters the direction of our life. The basilisk who slays by a look is no fable; it is a magical allegory. Generally speaking, it is bad for health to have enemies, and we can never brave with impunity the reprobation of anyone. Before opposing ourselves to a given force or current, we must be well assured that we possess the contrary force, or are with the stream of the contrary current; otherwise, we shall be crushed or struck down, and many sudden deaths have no other cause than this. The terrible visitations of Nadab and Abihu, of Osa, of Ananias and Saphira, were occasioned by electric currents of outraged convictions; the sufferings of the Ursulines of London, the nuns of Louviers and the convulsionaries of Jansenism, were identical in principle, and are explicable by the same occult natural laws. Had not Urban Grandier been immolated, one of two things would have occurred – either the possessed nuns would have died in frightful convulsions, or the phenomena of diabolical frenzy would have so gained in strength and influence, epidemically, that Grandier, notwithstanding his knowledge and his reason, would himself have become hallucinated, and to such a degree that he would have slandered himself, like the unhappy Gaufridy, or would otherwise have perished suddenly, with all the appalling characteristics of poisoning or of divine vengeance. In the eighteenth century the unfortunate poet Gilbert fell a victim to his audacity in braving the current of opinion and actual philosophical fanaticism which characterized his epoch. Guilty of philosophical treason, he died raving mad, possessed by the most incredible terrors, as if God himself had punished him for defending his cause out of season. As a fact, he perished by reason of a law of nature of which he could know nothing; he set himself against an electric current, and was struck down as by lightning. Had Marat not been assassinated by Charlotte Corday, he would have been destroyed infallibly by a reaction of public opinion. It was the execration of the honest which afflicted him with leprosy, and he would have had to succumb thereto. The reprobation excited by the massacre of St. Bartholomew was the sole cause of the atrocious disease and death of Charles IX, while, had not Henry IV been sustained by an immense popularity, which he owed to the projecting power or sympathetic force of his astral life, he would scarcely have outlived his conversion, but would have perished under the contempt of Protestants, combined with the suspicion and ill-will of Catholics. Unpopularity may be a proof of integrity and courage, but never of policy or prudence; the wounds inflicted by opinion are mortal for statesmen. We may recall the premature and violent end of many illustrious persons whom it would be inexpedient to mention here. Disgraces in public opinion may often be great injustices, but none the less they are invariably occasions of ill-success, and frequently of a death-sentence. In return, acts of injustice done to one individual can and should, if they rest unrepaired, cause the loss of an entire nation or of a whole society; this is what is called the cry of blood, for at the bottom of every injustice there is the germ of homicide. By reason of these terrible laws of solidarity, Christianity recommends so strongly the forgiveness of injuries and reconciliation. He who dies unforgiving casts himself dagger-armed into eternity and condemns himself to the horrors of an eternal murder. The efficacy of paternal or maternal blessings or curses is an invincible popular tradition and belief. As a fact, the closer the bonds which unite two persons, the more terrible are the consequences of hatred between them. The brand of Althæa burning the blood of Meleager is the mythological symbol of this terrible power. Let parents be ever on their guard, for no one can kindle hell in his own blood, or devote his own issue to misfortune, without being himself burnt and made wretched. To pardon is never a crime, but to curse is always a danger and an evil action.

9 ט I

INITIATION

JESOD BONUM

THE initiate is he who possesses the lamp of Trismegistus, the mantle of Apollonius, and the staff of the patriarchs. The lamp of Trismegistus is reason illuminated by science; the mantle of Apollonius is full and complete self-possession, which isolates the sage from blind tendencies; and the staff of the patriarchs is the help of the secret and everlasting forces of Nature. The lamp of Trismegistus enlightens present, past and future, lays bare the conscience of men, and manifests the inmost recesses of the female heart. The lamp burns with a triple flame, the mantle is thrice-folded, and the staff is divided into three parts.

The number nine is that of divine reflections; it expresses the divine idea in all its abstract power, but it also signifies extravagance in belief, and hence superstition and idolatry. For this reason Hermes made it the number of initiation, because the initiate reigns over superstition and by superstition, and alone can advance through the darkness, leaning on his staff, enveloped in his mantle, and lighted by his lamp. Reason has been given to all men, but all do not know how to make use of it; it is a science to be acquired. Liberty is offered to all, but not all can be free; it is a right that must be earned. Force is for all, but all do not know how to rest upon it; it is a power that must be seized. We attain nothing without more than one effort. The destiny of man is that he should enrich himself with what he gains, and that he should afterwards have, like God, the glory and pleasure of dispensing it.

Magic was called formerly the sacerdotal art and the royal art, because initiation gave empire over souls to the sage, and the adroitness for ruling wills. Divination is also one of the privileges of the initiate; now, divination is simply the knowledge of effects contained in causes and science applied to the facts of the universal dogma of analogy. Human acts are not alone written in the astral light; their traces are left upon the face, they modify mien and carriage, they change the tone of the voice. Thus, every man bears about him the history of his life, which is legible for the initiate. Now, the future is ever the consequence of the past, and unexpected circumstances do not appreciably alter results reasonably calculated. The destiny of each man can be therefore foretold him. An entire existence may be judged by a single movement; one piece of awkwardness may be the presage of a long chain of misfortunes. Cæsar was assassinated because he was ashamed of being bald; Napoleon ended his days at St. Helena because he admired the poems of Ossian; Louis Philippe abdicated the throne as he did because he carried an umbrella. These are paradoxes for the vulgar, who cannot grasp the occult relations of things, but they are causes for the adept, who understands all and is surprised at nothing.

Initiation is a preservative against the false lights of mysticism; it equips human reason with its relative value and proportional infallibility, connecting it with supreme reason by the chain of analogies. Hence the initiate knows no doubtful hopes, no absurd fears, because he has no irrational beliefs; he is acquainted with the extent of his power, and he can dare without danger. For him, therefore, to dare is to be able. Here, then, is a new interpretation of his attributes; his lamp represents learning, the mantle which enwraps him his discretion, and his staff is the emblem of his strength and daring. He knows, he dares and is silent. He knows the secrets of the future, he dares in the present, and he is silent on the past. He knows the failings of the human heart; he dares make use of them to achieve his work; and he is silent as to his purposes. He knows the principle of all symbolisms and of all religions; he dares to practice or abstain from them without hypocrisy and without impiety; and he is silent upon the one dogma of supreme initiation. He knows the existence and nature of the great magical agent; he dares perform the acts and give utterance to the words which make it subject to human will, and he is silent upon the mysteries of the great arcanum.

So, may you find him often melancholy, never dejected or despairing; often poor, never abject or miserable; persecuted often, never disheartened or conquered. He remembers the bereavement and murder of Orpheus, the exile and lonely death of Moses, the martyrdom of the prophets, the tortures of Apollonius, the Cross of the Savior. He knows the desolation in which Agrippa died, whose memory is even now slandered; he knows what labors overcame the great Paracelsus, and all that Raymond Lully was condemned to undergo that he might finish by a violent death. He remembers Swedenborg simulating madness and even losing reason in order to excuse his science; St. Martin and his hidden life; Cagliostro, who perished forsaken in the cells of the Inquisition; Cazotte, who ascended the scaffold. Inheritor of so many victims, he does not dare the less, but he understands better the necessity for silence. Let us follow his example; let us learn diligently; when we know, let us have courage, and let us be silent.

10 ׳ K

THE KABBALAH

MALCHUTH PRINCIPIUM PHALLUS

ALL religions have preserved the remembrance of a primitive book, written in types, by the sages of the earliest ages of the world; simplified and vulgarized in later days, its symbols furnished letters to the art of Writing, characters to the Word, and to occult Philosophy its mysterious signs and pantacles. This book, attributed by the Hebrews to Enoch, seventh master of the world after Adam; by the Egyptians to Hermes Trismegistus; by the Greeks to Cadmus, the mysterious builder of the Holy City; this book was the symbolical summary of primitive tradition, called subsequently Kabbalah or Cabala, meaning reception. The tradition in question rests altogether on the one dogma of Magic: the visible is for us the proportional measure of the invisible. Now the ancients, observing that equilibrium is the universal law in physics, consequent on the apparent opposition of two forces, argued from physical to metaphysical equilibrium, and maintained that in God, that is, in the prime living and active cause, there must be recognized two properties which are necessary one to another –

stability and motion, necessity and liberty, rational order and volitional autonomy, justice and love, whence also severity and mercy. Now, these two attributes were personified, so to speak, by the Kabbalistic Jews under the names of Geburah and Chesed. Above Geburah and Chesed abides the supreme crown, the equilibrating power, principle of the world or equilibrated kingdom, which we find mentioned under the name of Malchuth in the occult and kabbalistic versicle of the *Pater-noster* to which we have already referred. But Geburah and Chesed, maintained in equilibrium by the crown above and the kingdom below, constitute two principles, which may be considered from an abstract point of view, or in their realization. In their abstract or idealized sense, they take the higher names of *Chochmah*, Wisdom, and *Binah*, Intelligence. Their realization is stability and progress, that is, eternity and victory – *Hod* and *Netsah*.

Such, according to the Kabbalah, is the groundwork of all religions and all sciences – a triple triangle and a circle, the notion of the triad explained by the balance multiplied by itself in the domains of the ideal, then the realization of this conception in forms. Now, the ancients attached the first notions of this simple and impressive theology to the very idea of numbers, and qualified the figure of the first decade after the following manner:

1. *Kether*. The Crown, the equilibrating power.
2. *Chokmah*. Wisdom, equilibrated in its unchangeable order by the initiative of intelligence.
3. *Binah*. Active intelligence, equilibrated by Wisdom.
4. *Chesed*. Mercy, which is wisdom in its secondary conception, ever benevolent because it is strong.
5. *Geburah*. Austerity, necessitated by Wisdom itself, and by

goodwill. To permit evil is to hinder good.

6. *Tiphereth*. Beauty, the luminous conception of equilibrium in forms, intermediary between the Crown and the Kingdom, mediating principle between Creator and creation (a sublime conception of poetry and its sovereign priesthood!)

7. *Netsah*. Victory, that is, eternal triumph of intelligence and justice.

8. *Hod*. Eternity of the conquests achieved by mind over matter, active over passive, life over death.

9. *Jesod*. The Foundation, that is, the basis of all belief and all truth – the ABSOLUTE in philosophy.

10. *Malkuth*. The Kingdom is the universe, entire creation, the work and mirror of God, the proof of supreme reason, the formal consequence which compels us to have recourse to virtual premises, the enigma which has God for its answer – supreme and absolute reason.

These ten primary notions attached to the ten first characters of the primitive alphabet, signifying both principles and numbers, are called the ten Sephiroth by the masters in Kabbalah. The sacred tetragram, drawn in the following manner, indicates the number, source and correspondence of the divine names. To this name of Jotchavah, written by these four-and-twenty signs, crowned with a triple flower of light, must be referred the twenty-four thrones of heaven and the twenty-four crowned elders in the Apocalypse. In the Kabbalah the occult principle is called the Elder, and this principle, multiplied, and, as it were, reflected, in secondary causes, creates images of itself – that is to say, so many elders as there are diverse conceptions of its unique essence. These images, less perfect in proportion as they are removed farther from their source, project upon the darkness an ultimate reflection or glimmer, representing a horrible and deformed elder, who is vulgarly termed the devil. Hence an initiate has been bold enough to say: "The devil is God, as understood by the wicked"; while another has added, in words more bizarre but no less energetic: "The devil is composed of God's ruins." We may sum up and explain these strikingly novel definitions by remarking that in symbolism itself the demon is an angel cast out of heaven for having sought to usurp divinity. This belongs to the allegorical language of prophets and makers of legends. Philosophically speaking, the devil is a human idea of divinity, which has been surpassed and dispossessed of heaven by the progress of science and reason. Among primitive Oriental peoples, Moloch, Adramelek, Baal, were personifications of the one God, dishonored by barbarous attributes. The god of the Jansenists, creating hell for the majority of human beings, and delighting in the eternal tortures of those he was unwilling to save, is a conception even more barbarous than that of Moloch; hence the god of the Jansenists is already a veritable Satan, fallen from heaven, in the sight of every wise and enlightened Christian.

In the multiplication of the Divine Names the Kabalists have connected them all, either with the unity of the Tetragram, the figure of the triad, or the sephirotic scale of the decad. They arrange the scale of the Divine Names and numbers in a triangle, which may be presented in Roman characters as follows:

```
        J
       JA
       SDI
      JEHV
      ELOIM
      SABAOT
      ARARITA
     ELVEDAAT
     ELIM GIBOR
     ELIM SABAOT
```

The sum of all these divine names formed from the one tetragram is a basis of the Hebrew Ritual, and constitutes the occult force which the kabbalistic rabbins invoke under the title of Semhamphoras.

We have now to concern ourselves with the Tarot from the kabbalistic point of view, and have already indicated the occult source of the name. This hieroglyphic book is composed of a kabbalistic alphabet, and of a wheel or circle of four decades, distinguished by four symbolical and typical figures, each having for its radius a scale of four progressive figures, which represent Humanity: man, woman, youth, child – master, mistress, knight, esquire. The twenty-two figures of the alphabet represent, in the first place, the thirteen dogmas, and secondly, the nine beliefs authorized by that Jewish religion which is so strong and so firmly established in the highest reason.

Here follows the religious and kabbalistic key of The Tarot, formulated in technical verses after the mode of the ancient lawgivers:

1	א	A conscious, active cause in all we see,
2	ב	And number proves the living unity.
3	ג	No bound hath He who doth the whole contain,
4	ד	But, all preceding, fills life's vast domain.
5	ה	Sole worthy worship, He, the only Lord,
6	ו	Doth his true doctrine to clean hearts accord.
7	ז	But since faith's works a single pontiff need,
8	ח	One law have we, and at one altar plead;
9	ט	Eternal God for aye their base upholds.
10	י	Heaven and man's day alike his rule enfolds.
11	כ	In mercy rich, in retribution strong,
12	ל	His people's King he will upraise ere long.
13	מ	The tomb gives entrance to the promised land, Death only ends; life's vistas still expand.

These doctrines sacred, pure and steadfast shine;

And thus we close our number's scale divine.

14	נ	Good angels all things temper and assuage,
15	ס	While evil spirits burst with wrath and rage.
16	ע	God doth the lightning rule, the flame subdue.
17	פ	His word controls both Vesper and her dew.
18	צ	He makes the moon our watchman through the night.
19	ק	And by his sun renews the world in light.
20	ר	When dust to dust returns, His breath can call
20 or 21	ש	Life from the tomb which is the fate of all.
21 or 22	ת	His crown illuminates the mercy seat,

And glorifies the cherubs at His feet.

By the help of this purely dogmatic explanation we shall already understand the kabbalistic alphabet of the Tarot. Thus, Figure I, entitled the Buffoon, represents the active principle in the economy of divine and human autotelia[15]. Figure II, vulgarly called Pope Joan,

---

[15] Spelled as in the original edition. -pnw

represents dogmatic unity based upon numbers, and is the personification of the Kabbalah or the Gnosis. Figure III represents divine Spirituality under the emblem of a winged woman, holding in one hand the apocalyptic eagle, and in the other the world suspended from the end of her scepter. The other emblems are equally clear, and can be explained as easily as the first. Turning now to the four suits, namely, Clubs, Cups, Swords, and Circles or Pantacles, commonly called *Deniers* – all these are hieroglyphics of the tetragram. Thus, the Club is the Egyptian Phallus or Hebrew *jod*; the Cup is the cteïs or primitive *he*; the Sword is the conjunction of both, or the lingam, represented in Hebrew preceding the captivity by *vau*; while the Circle or Pantacle, image of the world, is the *he* final of the divine name. Now let us take a Tarot and combine all its emblems one by one into the Wheel or ROTA of William Postel; let us group the four aces, the four twos, and so on, together; we shall then have ten packs of cards giving the hieroglyphic interpretation of the triangle of divine names on the scale of the denary, as previously tabulated. By referring each number to its corresponding Sephira, we may then read them off as follows:

יהוה

Four signs present the name of every name.

1   KETHER.

The four Aces.

Four brilliant beams adorn His crown of flame.

2   CHOKMAH.

The four Twos.

Four rivers ever from his wisdom flow.

3   BINAH.

The four Threes.

Four proofs of his intelligence we know.

4   CHESED.

The four Fours.

Four benefactions from His mercy come.

5   GEBURAH.

The four Fives.

Four times four sins avenged his justice sum.

6   TIPHERETH.

The four Sixes.

Four rays unclouded make his beauty known.

7   NETSAH.

The four Sevens.

Four times his conquest shall in song be shown.

8   HOD.

The four Eights.

Four times he triumphs on the timeless plane.

9   JESOD.

The four Nines.

Foundations four his great white throne maintain.

10   MALCHUTH.

The four Tens.

One fourfold kingdom owns his endless sway,
As from his crown there streams a fourfold ray.

By this simple arrangement the kabbalistic meaning of each card is exhibited. For example, the five of clubs rigorously signifies Geburah of Jod, that is, the justice of the creator or the wrath of man; the seven of cups signifies the victory of mercy or the triumph of woman; the eight of swords signifies conflict or eternal equilibrium; and so of the others. We can thus understand how the ancient pontiffs proceeded to make the oracle speak. The chance dealing of the lamens produced invariably a fresh kabbalistic meaning, exactly true in its combinations, which alone were fortuitous; and, seeing that the faith of the ancients attributed nothing to chance, they read the answers of Providence in the oracles of the Tarot, which were called Theraph or Theraphim by the Hebrews, as the erudite kabbalist Gaffarel, one of the magicians employed by Cardinal Richelieu, was the first to perceive.

As to the figures, a final couplet will suffice to explain them:

KING, QUEEN, KNIGHT, ESQUIRE.
The married pair, the youth, the child, the race;
Thy path by these to Unity retrace.

At the end of the *Ritual* we shall provide further details, together with full documents, concerning the marvelous Tarot book, which of all books the most primitive, the key of prophecies and dogmas, in a word, the inspiration of inspired works, a fact which has remained unperceived equally by the science of Court de Gebelin and by the extraordinary intuitions of Eteilla or Alliette.

The ten Sephiroth and the twenty-two Tarots form what the kabbalists term the thirty-two paths of absolute science. With regard to particular sciences, they distinguish them into fifty chapters, which they call the fifty gates – among Orientals the word gate signifies government or authority. The rabbins also divided the Kabbalah into Bereschit, or universal Genesis, and Mercavah, or the chariot of Ezekiel; then by means of a dual interpretation of the kabbalistic alphabets, they formed two sciences, called Gematria and Temurah, and so composed the notary art, which is fundamentally the complete science of the Tarot signs and their complex and varied application to the divination of all secrets, whether of philosophy, nature, or the future itself. We shall recur in our twentieth chapter to this work.

11 ב K

THE MAGIC CHAIN
MANUS FORCE

THE great magical agent, by us termed the astral light, by others the soul of the earth, and designated by old chemists under the names of Azoth and Magnesia, this occult, unique and indubitable force, is the key of all empire, the secret of all power; it is the winged dragon of Medea, the serpent of the Edenic mystery; it is the universal glass of visions, the bond of sympathies, the source of love, prophecy and glory. To know how to avail one's self of this agent is to be the trustee of God's own power; all real, effective Magic, all occult force is there, and its demonstration is the sole end of all genuine books of science. To possess one's self of the great magical agent there are two operations necessary – to concentrate and project, or, in other words, to fix and to move. Fixity has been provided as the basis and guarantee of movement by the Author of all things: the magus must go to work in like manner.

It is said that enthusiasm is contagious – and why? Because it cannot be produced in the absence of collective faith. Faith produces faith; to believe is to have a reason for willing; to will with reason is to will with power – not, I say, with an infinite, but with an indefinite power. What operates in the intellectual and moral world obtains still more in the physical, and when Archimedes was in want of a lever to move the world, what he sought was simply the great magical arcanum. One arm of the androgyne figure of Henry Khunrath bore the word COAGULA and the other SOLVE. To collect and diffuse are nature's two words – but after what manner can we accomplish these operations with the astral light or soul of the world? Concentration is by isolation and distribution by the magical chain. Isolation consists in absolute independence for thought, complete liberty for the heart and perfect continence for the senses. Every man who is possessed by prejudices and fears, every passionate person who is slave of his passions, is incapable of concentrating or coagulating, according to the expression of Khunrath, the astral light or soul of the earth. All true adepts have been independent even amidst torture, sober and chaste till death. The explanation of such anomaly is this – in order to dispose of a force, you must not be surprised by this force in a way that it may dispose of you. But then, cry out those who seek only in magic for a method of inordinately satisfying the lusts of nature, what good is power which must not be used for our own satisfaction? Unhappy creatures who ask, if I told you, how could you grasp it? Are pearls nothing because they are worthless to the horde of Epicurus? Did not Curtius prefer the government of those who had gold rather than its possession by himself? Must we not be something removed from the common man when we almost pretend to be God? Moreover, I grieve to deject or discourage you, but I am not devising the transcendental sciences; I teach them, defining their immutable necessities in the presentation of their primary and most inexorable conditions. Pythagoras was a free, sober and chaste man; Apollonius of Tyana and Julius Caesar were both of repellent austerity; the sex of Paracelsus was suspected, so foreign was he to the weakness of love; Raymond Lully carried the severity of life to the most exalted point of asceticism; Jerome Cardan exaggerated the practice of fasting till he nearly perished of starvation, if we may accept tradition; Agrippa, poor and buffeted from town to town, almost died of misery rather than yield to the caprice of a princess who insulted the liberty of science. What then made the happiness of these men? The knowledge of great secrets and the consciousness of power. It was sufficient for those exalted souls. Must one be like unto them in order to know what they knew? Assuredly not, and the existence of this book is perhaps a case in point; but in order to do what they did, it is absolutely necessary to take the means which they took. Yet what did they actually accomplish? They astonished and subdued the world; they reigned more truly than kings. Magic is an instrument of divine goodness or demoniac pride, but it is the annihilation of earthly joys and the pleasures of mortal life. Why study it? ask the luxurious. Merely to know it, and possibly after to learn mistrust of stupid unbelief or puerile credulity. Men of pleasure, and half of these I count for so many women, is not gratified curiosity highly pleasurable? Read therefore without fear, you will not be magicians against your will. Readiness for absolute renunciation is, moreover, necessary only in order to establish universal currents and transform the face of the world; there are relative magical operations, limited to a certain circle, which do not need such heroic virtues. We can act upon passions by passions, determine sympathies or antipathies, hurt even and heal, without possessing the omnipotence of the magus; in this case, however, we must realize the risk of a reaction in proportion to the action, and to such risk we may fall easily a victim. All this will be explained in our Ritual.

To make the Magic Chain is to establish a magnetic current which becomes stronger in proportion to the extent of the chain. We shall see in the Ritual how such currents can be produced, and what are the various modes of forming the chain. Mesmer's trough was an exceedingly imperfect magic chain; several great circles of illuminati in different northern countries possess more potent chains. Even that association of Catholic priests, celebrated for their occult power and their unpopularity, is established upon the plan and follows the conditions of the most potent magical chains, and herein is the secret of their force, which they attribute solely to the grace or will of God, a vulgar and cheap solution for every mystery of power in influence or attraction. In the Ritual it will be our task to estimate the sequence of truly magical ceremonies and evocations which constitute the great work of vocation under the name of the Exercises of St. Ignatius.

All enthusiasm propagated in a society by a series of communications and practices in common produces a magnetic current, and continues or increases by the current. The action of the current is to carry away and often to exalt beyond measure persons who are impressionable and weak, nervous organizations, temperaments inclined to hysteria or hallucination. Such people soon become powerful vehicles of magical force and efficiently project the astral light in the direction of the current itself; opposition at such a time to the manifestations of the force is, to some extent, a struggle with fatality. When the youthful Pharisee Saul, or Schôl, threw himself, with all the fanaticism and all the determination of a sectarian, across the aggressive line of Christianity, he unconsciously placed himself at the mercy of a power against which he thought to prevail, and hence he was struck down by a formidable magnetic flash, doubtless the more instantaneous by reason of the combined effect of cerebral congestion and sunstroke.

The conversion of the young Israelite, Alphonus of Ratisbonne, is a contemporary fact which is absolutely of the same nature. We are acquainted with a sect of enthusiasts whom it is common to deride at a distance, and to join, despite one's self, as soon as they are approached, even with a hostile intention. I will go further, and affirm that magical circles and magnetic currents establish themselves, and have an influence, according to fatal laws, upon those on whom they can act. Each one of us is drawn within a circle of relations which constitutes his world, and to the influence of which he is made subject. The lawgiver of the French Revolution, that man whom the most spiritual nation in the whole world acknowledged as the incarnation of human reason, Jean Jacques Rousseau, was drawn into the most lamentable action of his life, the desertion of his children, by the magnetic influence of a libertine circle and a magical current of *table-d'hôte*. He describes it simply and ingenuously in his Confessions, but it is a fact which has remained unobserved. Great circles very often make great men, and *vice-versâ*. There are no unrecognized geniuses, there are *eccentric* men, and the term would seem to have been invented by an adept. The man who is eccentric in his genius is one who attempts to form a circle by combating the central attractive force of established chains and currents. It is his destiny to be broken or to succeed. Now, what is the twofold condition of success in such a case? A central point of stability and a persistent circular action of initiative. The man of genius is one who has discovered a real law, and is possessed thereby of an invincible, active, and grinding power. He may die in the midst of his work, but that which he has willed comes to pass, in spite of his death, and is indeed often ensured thereby, because death is a veritable assumption for genius. "When I shall be lifted up from the earth," said the greatest of the initiators, "I will draw all things after me."

The law of magnetic currents is that of the movement of the astral light itself, which is always double, and augments in an opposite sense. A great action invariably paves the way for a reaction of equal magnitude, and the secret of phenomenal successes consists entirely in the foreknowledge of reactions. Thus, did Chateaubriand, penetrated with disgust at the saturnalia of the revolution, foresee and prepare the immense vogue of his "Genius of Christianity." To oppose one's self to a current at the beginning of its revolution is to court being destroyed by that current, like the great and unfortunate Emperor Julian; to oppose one's self to a current which has run its round is to take the lead of a contrary current. The great man is he who comes seasonably and knows how to innovate opportunely. In the days of the apostles, Voltaire would have found no echo for his utterances, and might have been merely an ingenious parasite at the banquets of Trimalcyon. Now, at the epoch wherein we live, everything is ripe for a fresh outburst of evangelical zeal and Christian self-devotion, precisely by reason of the prevailing general disillusion, egoistic positivism and public cynicism of the coarsest interests. The success of certain books and the mystical tendencies of minds are unequivocal symptoms of this widespread disposition. We restore and we build churches only to realize more keenly that we are void of belief, only to long the more for it; once more does the whole world await its Messiah, and he cannot tarry in his coming. Let a man, for example, come forward, who by rank or by fortune is placed in an exalted position – a pope, a king, even a Jewish millionaire – and let this man publicly and solemnly sacrifice all his material interests for the weal of humanity; let him make himself the savior of the poor, the disseminator, and even the victim, of doctrines of renunciation and charity, and he will draw round him an immense following; he will accomplish a complete moral revolution in the world. But the high place is before all things necessary for such a personage, because, in these days of meanness and trickery, any Word issuing from the lower ranks is suspected of interested ambition and imposture. Ye, then, who are nothing, ye who possess nothing, aspire not to be apostles or messiahs. If you have faith, and would act in accordance therewith, get possession, in the first place, of the means of action, which are the influence of rank and the prestige of fortune. In olden times gold was manufactured by science; nowadays science must be remade by gold. We have fixed the volatile, and we must now volatilize the fixed – in other words, we have materialized spirit, and we must now spiritualize matter. The most sublime utterance now passes unheeded if it goes forth without the guarantee of a name – that is to say, of a success which represents a material value. What is the worth of a manuscript? That of the author's signature among the booksellers? That established reputation known as Alexandre Dumas et Cie represents one of the literary guarantees of our time, but the house of Dumas is in repute only for the romances which are its exclusive productions. Let Dumas devise a magnificent Utopia, or discover a splendid solution of the religious problem, and no one will take them seriously, despite the European celebrity of the Panurge of modern literature. We are in the age of acquired positions, where everyone is appraised according to his social and commercial standing. Unlimited freedom of speech has produced such a strife of words that no one inquires what is said, but who has said it. If it be Rothschild, his Holiness Pius the Ninth, or even Monseigneur Dupanloup, it is something; but if it be Tartempion, it is nothing, were he even – which is possible, after all – an unrecognized prodigy of genius, knowledge and good sense. Hence to those who would say to me: If you possess the secret of great successes, and of a force which can transform the world, why do you not make use of them? I would answer: This knowledge has come to me too late for myself, and I have spent over its acquisition the time and the resources which might have enabled me to apply it; I offer it to those who are in a position to avail themselves of it. Illustrious men, rich men, great ones of this world, who are dissatisfied with that which you have, who are conscious of a nobler and larger ambition, will you be fathers of a new world, kings of a rejuvenated civilization? A poor and obscure scholar has found the lever of Archimedes, and he offers it to you for the good of humanity alone, asking nothing whatsoever in exchange.

The phenomena which quite recently have perturbed America and Europe, those of table-turning and fluidic manifestations, are simply magnetic currents at the beginning of their formation, appeals on the part of nature inviting us, for the good of humanity, to re-establish great sympathetic and religious chains. As a fact, stagnation in the astral light would mean death to the human race, and torpor in this secret agent has already been manifested by alarming symptoms of decomposition and death. For example, cholera-morbus, the potato disease, and the blight on the grape, are traceable solely to this cause, as the two young shepherds of la Salette saw darkly and symbolically in their dream. The unlooked-for credit which awaited their narrative, and the vast concourse of pilgrims attracted by a statement so singular and at the same time so vague as that of these two children without instruction and almost without morality, are proofs of the magnetic reality of the fact, and the fluidic tendency of the earth itself to operate the cure of its inhabitants. Superstitions are instinctive, and all that is instinctive is founded in the very nature of things, to which fact the sceptics of all times have given insufficient attention. We attribute, then, all the strange phenomena of table-turning to the universal magnetic agent in search of a chain of enthusiasms with a view to the formation of fresh currents. The force of itself is blind, but can be directed by the will of man, and is influenced by prevailing opinions. This universal fluid – if we decide to regard it as a fluid – being the common medium of all nervous organisms, and the vehicle of all sensitive vibrations, establishes an actual physical solidarity between impressionable persons, and transmits from one to another the impressions of imagination and of thought. The movement of the inert object, determined by the undulations of the universal agent, obeys the ruling impression and reproduces in its revelations at one time all the lucidity of the most wonderful visions, but at another all the eccentricity and falsehood of the most vague and incoherent dreams. The blows resounding on furniture, the clattering of dishes, the auto-playing of musical instruments, are illusions produced by the same cause. The miracles of the convulsionaries of Saint Médard were of the same order, and seemed frequently to suspend the laws of nature. On the one hand, exaggeration produced by fascination, which is the special quality of intoxication occasioned by congestions of the astral light; on the other, actual oscillations or movements impressed upon inert matter by the subtle and universal agent of motion and life. Such is the sole foundation of these occurrences which look so marvelous, as we may demonstrate easily at will by reproducing, in accordance with rules laid down in the Ritual, the most astounding of these phenomena, establishing, as can be done quite simply, the absence of trickery, hallucination, or error.

It has frequently happened to me after experiments in the magic chain, performed with persons devoid of good intention or sympathy, that I have been awakened with a start in the night by truly alarming impressions and sensations. On one such occasion I felt vividly the pressure of an unknown hand attempting to strangle me; I rose up, lighted my lamp, and set calmly to work, seeking to profit by my wakefulness and to drive away the phantoms of sleep. The books about me were moved with much noise, papers were disturbed and rubbed one against another, timber creaked as if on the point of splitting, and heavy blows resounded on the ceiling. With curiosity but also with tranquility I observed all these phenomena, which would not have been less wonderful had they been only the product of my imagination, so real did they seem. For the rest, I may state that I was in no sense frightened, and during this occurrence I was engaged upon something quite foreign to the occult sciences. By the repetition of similar phenomena, I was led to attempt an experience of evocation, assisted by the magical ceremonies of the ancients, when I obtained truly astounding results, which will be set forth in the thirteenth chapter of this work.

12 ל L

THE GREAT WORK

DISCITE CRUX

THE great work is, before all things, the creation of man by himself, that is to say, the full and entire conquest of his faculties and his future; it is especially the perfect emancipation of his will, assuring him universal dominion over Azoth and the domain of Magnesia, in other words, full power over the universal magical agent. This agent, disguised by the ancient philosophers under the name of the first matter, determines the forms of modifiable substance, and we can really arrive by means of it at metallic transmutation and the universal medicine. This is not a hypothesis, it is a scientific fact already established and rigorously demonstrable. Nicholas Flamel and Raymond Lully, both of them poor, indubitably distributed immense riches. Agrippa never proceeded beyond the first part of the great work, and he died in the ordeal, fighting to possess himself and to fix his independence.

Now, there are two Hermetic operations, the one spiritual, the other material, and these are mutually dependent. For the rest, all Hermetic science is contained in the doctrine of Hermes, which is said to have been originally inscribed upon an emerald tablet. Its first articles have been already expounded, and those follow which are concerned with the operation of the Great Work:

"Thou shalt separate the earth from the fire, the subtle from the gross, gently, with great industry. It rises from earth to heaven, and again it descends to earth, and it receives the power of things above and of things below. By this means shalt thou obtain the glory of the whole world, and all darkness shall depart from thee. It is the strong power of every power, for it will overcome all that is subtle and penetrate all that is solid. Thus, was the world created."

To separate the subtle from the gross, in the first operation, which is wholly inward, is to liberate the soul from all prejudice and all vice, which is accomplished by the use of the philosophical salt, that is to say, wisdom; of mercury, that is, personal skill and application; finally, of sulphur, representing vital energy and fire of will. By these are we enabled to change into spiritual gold things which are of all least precious, even the refuse of the earth. In this sense we must interpret the parables of the choir of philosophers, Bernard Trevisan, Basil Valentine, Mary the Egyptian and other prophets of alchemy; but in their works, as in the great work, we must adroitly separate the subtle from the gross, the mystical from the positive, allegory from theory. If we would read them with profit and understanding, we must take them first of all as allegorical in their entirety, and then descend from allegories to realities by the way of the correspondences or analogies indicated in the one dogma: That which is above is proportional to that which is below, and reciprocally. The word ART when reversed or read after the manner of sacred and primitive characters from right to left, gives three initials which express the different grades of the great work. T signifies triad, theory and travail; R, realization; A, adaptation. In the twelfth chapter of the Ritual, we shall give the processes for adaptation, in use among the great masters, especially that which is contained in the Hermetic Citadel of Henry Khunrath. We may refer our readers also to an admirable treatise attributed to Hermes Trismegistus and entitled *Minerva Mundi*. It is found only in certain editions of Hermes, and contains, beneath allegories full of profundity and poetry, the doctrine of individual self-creation, or the creative law consequent on the harmony between two forces, which are termed fixed and volatile by alchemists, and are necessity and liberty in the absolute order. The diversity of the forms which abound in nature is explained, in this treatise, by the diversity of spirits, and monstrosities by the divergence of efforts; its reading and assimilation are indispensable for all adepts who would fathom the mysteries of nature and devote themselves seriously to the search after the great work.

When the masters in alchemy say that a short time and little money are needed to accomplish the works of science, above all when they affirm that one vessel is alone needed, when they speak of the great and unique athanor, which all can use, which is ready to each man's hand, which all possess without knowing it, they allude to philosophical and moral alchemy. As a fact, a strong and resolute will can arrive in a short time at absolute independence, and we are all in possession of the chemical instrument, the great and sole athanor which answers for the separation of the subtle from the gross and the fixed from the volatile. This instrument, complete as the world and precise as mathematics, is represented by the sages under the emblem of the pentagram or five-pointed star, which is the absolute sign of human intelligence. I will follow the example of the wise by forbearing to name it; it is too easy to guess it.

The Tarot symbol which corresponds to this chapter was misconstrued by Court de Gebelin and Etteilla, who regarded it as a blunder of a German card maker. It represents a man with his hands bound behind him, having two bags of silver attached to the armpits, and being suspended by one foot from a gibbet formed by the trunks of two trees, each with a root of six lopped branches, and by a crosspiece, thus completing the figure of the Hebrew *tau* ת; the legs of the victim are crossed, and his head and elbows form a triangle. Now, the triangle surmounted by a cross signifies, in alchemy, the end and perfection of the great work, a meaning which is identical with that of the letter *tau*, the last of the sacred alphabet. The hanged man is, consequently, the adept, bound by his engagements, and spiritualized, that is, having his feet turned towards heaven; it is also the antique Prometheus, expiating by everlasting torture the penalty of his glorious theft; vulgarly, he is the traitor Judas, and his punishment threatens betrayers of the great arcanum. Finally, for Kabbalistic Jews, the hanged man, who corresponds to their twelfth doctrine, that of the promised Messiah, is a protestation against the Savior acknowledged by Christians, and they seem to say unto Him still: How canst thou save others, since thou couldst not save thyself?

In the *Sepher-Toldos-Jeschu*, an anti-Christian rabbinical compilation, there occurs a singular parable. Jeshu, says the rabbinical author of the legend, was travelling with Simon-Barjona and Judas Iscariot. Late and weary they came to a lonely house, and, being very hungry, could find nothing to eat except an exceedingly lean gosling. It was insufficient for three persons, and to divide it would be to sharpen without satisfying hunger. They agreed to draw lots, but as they were heavy with sleep: "Let us first of all slumber," said Jeshu, "whilst the supper is preparing; when we wake we will tell our dreams, and he who has had the most beautiful dream shall have the whole gosling to his own share." So it was arranged; they slept and they woke. As for me, said St. Peter, I dreamed that I was the vicar of God. And I, said Jeshu, that I was God himself. For me, said Judas hypocritically, I dreamed that, being in somnambulism, I arose, went softly downstairs, took the gosling from the spit, and ate it. Thereupon they also went down, but the gosling had vanished altogether. Judas had a waking dream.

This anecdote is given, not in the text of the Sepher-Toldos-Jeschu itself, but in the rabbinical commentaries on that work. The legend is a protest of Jewish positivism against Christian mysticism. As a fact, while the faithful surrendered themselves to magnificent dreams, the proscribed Israelite, Judas of the Christian civilization, worked, sold, intrigued, became rich, possessed himself of this life's realities, till he became in a position to advance the means of existence to those very forms of worship which had so long outlawed him. The ancient adorers of the ark remained true to the cultus of the strong-box; the Exchange is now their temple, and thence they govern the Christian world. The laugh is indeed with Judas, who can congratulate himself upon not having slept like St. Peter.

In archaic writings preceding the Captivity, the Hebrew *tau* was cruciform, which further confirms our interpretation of the twelfth plate of the Kabbalistic Tarot. The cross, which produces four triangles, is also the sacred sign of the duodenary, and on this account it was called the Key of Heaven by the Egyptians. So Etteilla, confused by his protracted researches for the conciliation of the analogical necessities of this symbol with his own personal opinion, in which he was influenced by the erudite Court de Gebelin, placed in the hand of his upright hanged man, by him interpreted as Prudence, a Hermetic caduceus, formed by two serpents and a Greek *tau*. Seeing that he understood the necessity of the *tau* or Cross on the twelfth leaf of the Book of Thoth, he should have seen also the manifold and magnificent meaning of the Hermetic hanged man, the Prometheus of science, the living man who touches earth by his thought alone, whose firm ground is heaven, the free and immolated adept, the revealer menaced with death, the conjuration of Judaism against Christ, which seems to be an involuntary admission of the secret divinity of the Crucified, and lastly, the sign of the work accomplished, the cycle terminated, the intermediary *tau*, which resumes for the first time, before the final denary, the signs of the sacred alphabet.

13 ‎מ N

NECROMANCY

EX IPSIS MORS

WE have said that the images of persons and things are preserved in the astral light. Therein also can be evoked the forms of those who are in our world no longer, and by this means are accomplished those mysteries of necromancy which are so contested and at the same time so real. The Kabalists who have discoursed concerning the world of spirits have described simply what they have seen in their evocations. Éliphas Lévi Zahed,[16] who writes this book, has evoked, and he has seen. Let us state, in the first place, what the masters have written of their visions or their intuitions in that which they term the light of glory. We read in the Hebrew book concerning the *Revolution of Souls* that there are three classes of souls – the daughters of Adam, the daughters of angels and the daughters of sin. According to the same work, there are also three kinds of spirits – captive spirits, wandering spirits, and free spirits. Souls are sent forth in couples; at the same time certain souls of men are born widowed, for their spouses are held captive by Lilith and Naëmah, the queens of the stryges; they are souls

---

[16] These Hebrew names translated into French are Alphonse Louis Constant.

condemned to expiate the temerity of a celibate's vow. Hence, when a man renounces the love of women from his infancy, he makes the bride who was destined for him a slave to the demons of debauch. Souls grow and multiply in heaven, as bodies do upon earth. Immaculate souls are the daughters of the kisses of angels.

Nothing can enter heaven save that which comes from heaven. Hence, after death, the divine spirit which animated man returns alone to heaven and leaves two corpses, one upon earth, the other in the atmosphere; one terrestrial and elementary, the other aerial and sidereal, one already inert, the other still animated by the universal movement of the soul of the world, yet destined to die slowly, absorbed by the astral forces which produced it. The terrestrial body is visible; the other is unseen by the eyes of earthly and living bodies, nor can it be beheld except by the application of the astral light to the *translucid*, which conveys its impressions to the nervous system, and thus influences the organ of sight so as to make it perceive the forms which are preserved and the words which are written in the book of vital light.

When a man has lived well the astral body evaporates like a pure incense ascending towards the upper regions; but should he have lived in sin, his astral body, which holds him prisoner, still seeks the objects of its passions, and wishes to return to life. It torments the dreams of young girls, bathes in the steam of spilt blood, and floats about the places where the pleasures of its life elapsed; it watches over treasures which it possessed and buried; it expends itself in painful efforts to make fresh material organs and so live again. But the stars draw it up and absorb it; it feels its intelligence weaken, its memory gradually vanishes, all its being dissolves . . . . Its former vices rise up before it, assume monstrous shapes, and pursue it; they attack and devour it . . . . The unfortunate creature thus loses successively all the members which have ministered to his iniquities; then he dies a second time and for ever, because he loses his personality and his memory. Souls which are destined to live, but are not yet purified completely, remain captive for a longer or shorter period in the astral body, wherein they are burned by the odic light, which seeks to absorb and dissolve them. It is in order to escape from this body that suffering souls sometimes enter the bodies of the living and dwell therein in that state which Kabalists term embryonic. Now, it is these aerial bodies which are evoked by Necromancy. We enter into communion with larvæ, with dead or perishing substances, by this operation. The beings in question, for the most part, cannot speak except by the tingling of our ears produced by the nervous shock to which I have referred, and commonly they can reason only by reflecting our thoughts and our reveries. To behold these strange forms, we must put ourselves in abnormal condition akin to sleep or death, in other words, we must magnetize ourselves and enter into a kind of lucid and waking somnambulism. Then necromancy has real results, and then the evocations of Magic can produce actual visions. We have said that in the great magical agent, which is the astral light, there are preserved all impressions of things, all images formed either by rays or reflections; in this same light our visions come to us, and it is this which intoxicates the insane, and leads away their dormant judgement in pursuit of the most bizarre phantoms. To insure vision without illusion in this light, a powerful will must be with us to isolate reflections and attract rays only. To dream awake is to see in the astral light, and the orgies of the Sabbath, described by so many sorcerers in their criminal trials, came to them solely in this manner. The preparations and the substances used to obtain this result were often horrible, as we shall see in the Ritual, but the result itself was never doubtful. They beheld, they heard, they handled the most abominable, most fantastic, most impossible things. We shall return to this subject in our fifteenth chapter; at the present moment we are concerned only with evocations of the dead.

In the spring of the year 1854 I had undertaken a journey to London, that I might escape from internal disquietude and devote myself, without interruption, to science. I had letters of introduction to persons of eminence, who were anxious for revelations from the supernatural world. I made the acquaintance of several, and discovered in them, amidst much that was courteous, a depth of indifference or trifling. They asked me forthwith to work wonders, as if I were a charlatan, and I was somewhat discouraged, for, to speak frankly, far from being inclined to initiate others into the mysteries of Ceremonial Magic, I had myself shrunk all along from its illusions and weariness; moreover, such ceremonies necessitated an equipment which would be expensive and hard to collect. I buried myself, therefore, in the study of the transcendent Kabbalah, and concerned myself no further with English adepts, when, returning one day to my hotel, I found a note awaiting me. This note contained half of a card, divided transversely, on which I immediately recognized the seal of Solomon. It was accompanied by a small sheet of paper, on which these words were penciled: "Tomorrow, at three o'clock, in front of Westminster Abbey, the second half of this card will be given you." I kept this curious assignation. At the appointed spot I found a carriage drawn up, and as I held unaffectedly the fragment of card in my hand, a footman approached, making a sign as he did so, and then opened the door of the equipage. It contained a lady in black, wearing a thick veil; she motioned to me to take a seat beside her, showing me at the same time the other half of the card. The door closed, the carriage drove off, and, the lady raising her veil, I saw that my appointment was with an elderly person, having grey eyebrows and black eyes of unusual brilliance, strangely fixed in expression. "Sir," she began, with a strongly marked English accent, "I am aware that the law of secrecy is rigorous amongst adepts; a friend of Sir B— L—, who has seen you, knows that you have been asked for phenomena, and that you have refused to gratify such curiosity. You are possibly without the materials; I should like to show you a complete magical cabinet, but I must exact beforehand the most inviolable silence. If you will not give me this pledge upon your honor, I shall give orders for you to be driven to your home." I made the required promise, and faithfully keep it by divulging neither the name, position, nor abode of this lady, whom I soon recognized as an initiate, not exactly of the first order, but still of a most exalted grade. We had a number of long conversations, in the course of which she invariably insisted upon the necessity of practical experience to complete initiation. She showed me a collection of magical vestments and instruments, lent me some rare books, which I needed; in short, she determined me to attempt, at her house, the experiment of a complete evocation, for which I prepared during a period of twenty-one days, scrupulously observing the rules laid down in the thirteenth chapter of the Ritual.

The probation terminated on 24th of July: it was proposed to evoke the phantom of the divine Apollonius, and to question it upon two secrets, one which concerned myself, and one which interested the lady. She had counted on taking part in the evocation with a trustworthy person, but this person proved nervous at the last moment, and, as the triad or unity is indispensable for magical rites, I was left to my own resources. The cabinet prepared for the evocation was situated in a turret; it contained four concave mirrors, and a species of altar having a white marble top, encircled by a chain of magnetized iron. The sign of the pentagram, as given in the fifth chapter of this work, was carved and gilded on the white marble surface; it was drawn also in various colors upon a new white lambskin stretched beneath the altar. In the middle of the marble table there was a small copper chafing-dish, containing charcoal of alder and laurel wood; another chafing-dish was set before me on a tripod. I was clothed in a white garment, very similar to the vestments of our catholic priests, but longer and wider, and I wore upon my head a crown of vervain leaves, intertwined with a golden chain. I held a new sword in one hand, and in the other the Ritual. I kindled two fires with the required and prepared substances, and began reading the evocations of the Ritual in a voice at first low, but rising by degrees. The smoke spread, the flame caused the objects upon which it fell to waver, then it went out, the smoke still floating white and slow about the marble altar; I seemed to feel a kind of quaking of the earth, my ears tingled, my heart beat quickly. I heaped more twigs and perfumes on the chafing-dishes, and as the flame again burst up, I beheld distinctly, before the altar, the figure of a man of more than normal size, which dissolved and vanished away. I recommenced the evocations, and placed myself within a circle which I had drawn previously between the tripod and the altar. Thereupon the mirror which was behind the altar seemed to brighten in its depth, a wan form was outlined therein, which increased, and seemed to approach by degrees. Three times, and with closed eyes, I invoked Apollonius. When I again looked forth there was a man in front of me, wrapped from head to foot in a species of shroud, which seemed more grey than white; he was lean, melancholy and beardless, and did not altogether correspond to my preconceived notion of Apollonius. I experienced an abnormally cold sensation, and when I endeavored to question the phantom I could not articulate a syllable. I therefore placed my hand upon the sign of the pentagram, and pointed the sword at the figure, commanding it mentally to obey and not alarm me, in virtue of the said sign. The form thereupon became vague, and suddenly disappeared. I directed it to return, and presently felt, as it were, a breath close by me; something touched my hand which was holding the sword, and the arm became immediately benumbed as far as the elbow. I divined that the sword displeased the spirit, and I therefore placed its point downwards, close by me, within the circle. The human figure reappeared immediately, but I experienced such an intense weakness in all my limbs, and a swooning sensation came so quickly over me, that I made two steps to sit down, whereupon I fell into a profound lethargy, accompanied by dreams, of which I had only a confused recollection when I came again to myself. For several subsequent days my arm remained benumbed and painful. The apparition did not speak to me, but it seemed that the questions I had designed to ask answered themselves in my mind. To that of the lady an interior voice replied – Death! – it was concerning a man about whom she desired information. As for myself, I sought to know whether reconciliation and forgiveness were possible between two persons who occupied my thoughts, and the same inexorable echo within me answered – Dead!

I am stating facts as they occurred, but I would impose faith on no one. The consequence of this experience on myself was something inexplicable. I was no longer the same man; something of another world had passed into me; I was no longer either sad or cheerful, but I felt a singular attraction towards death, unaccompanied, however, by any suicidal tendency. I analyzed my experience carefully, and, notwithstanding a lively nervous repugnance, I twice repeated the same experiment, allowing some days to elapse between each; there was not, however, sufficient difference between the phenomena to warrant me in protracting a narrative which is perhaps already too long. But the net result of these two additional evocations was for me the revelation of two Kabbalistic secrets which might change, in a short space of time, the foundations and laws of society at large, if they came to be known generally.

Am I to conclude from all this that I really evoked, saw and touched the great Apollonius of Tyana? I am not so hallucinated as to affirm or so unserious as to believe it. The effect of the probations, the perfumes, the mirrors, the pantacles, is an actual drunkenness of the imagination, which must act powerfully upon a person otherwise nervous and impressionable. I do not explain the physical laws by which I saw and touched; I affirm solely that I did see and that I did touch, that I saw clearly and distinctly, apart from dreaming, and this is sufficient to establish the real efficacy of magical ceremonies. For the rest, I regard the practice as destructive and dangerous; if it became habitual, neither moral nor physical health would be able to withstand it. The elderly lady whom I have mentioned, and of whom I subsequently had reason to complain, was a case in point; despite her asseverations to the contrary, I have no doubt that she was addicted to necromancy and goëtia. She at times lost all self-control, at others yielded to senseless fits of passion, for which it was difficult to discover a cause. I left London without bidding her adieu, and I shall faithfully adhere to my engagement by giving no clue to her identity, which might connect her name with practices, pursued in all probability without the knowledge of her family, which I believe to be large and of very considerable position.

There are evocations of intelligence, evocations of love, and evocations of hate; but, once more, there is no proof whatsoever that spirits leave the higher spheres to communicate with us; the opposite, as a fact, is more probable. We evoke the memories which they have left in the astral light, or common reservoir of universal magnetism. It was in this light that the Emperor Julian once saw the gods manifest, looking old, ill and decrepit – fresh proof of the influence exercised by current and accredited opinions on the reflections of this same magical agent which makes our tables talk and answers by taps on the walls. After the evocation I have described, I re-read carefully the life of Apollonius, who is represented by historians as an ideal of antique beauty and elegance, and I then observed that towards the end of his life he was starved and tormented in prison. This circumstance, which may have remained in my memory without my being aware of it, possibly determined the unattractive form of my vision, which I regard solely as the voluntary dream of a waking man. I have seen two other persons, whom there is no occasion to name, both differing, as regards costume and appearance, from what I had expected. For the rest, I commend the greatest caution to all who propose devoting themselves to similar experiences; their result is intense exhaustion, and frequently a shock sufficient to occasion illness.

I must not conclude this chapter without mentioning the curious opinions of certain Kabbalists, who distinguish between apparent and real death, holding that the two are seldom simultaneous. In their idea, the majority of persons who are buried are still alive, while a number of others who are regarded as living are in reality dead. Incurable madness, for example, would be with them an incomplete but real death, leaving the earthly form under the purely instinctive control of the sidereal body. When the human soul suffers a greater blow than it can bear, it would thus become separated from the body, leaving the animal soul, or sidereal body, in its place, and these human remains would be to some extent less alive really than a mere animal. Dead persons of this kind are said to be recognized by the complete extinction of the moral and affectionate sense; they are neither bad nor good; they are dead. Such beings, who are poisonous fungi of the human race, absorb the life of living beings to their fullest possible extent, and this is why their proximity benumbs the soul and chills the heart. If such corpse-like creatures really existed, they would realize all that was recounted in former times about brucalaques and vampires. Now, are there not certain persons in whose presence one feels less intelligent, less good, sometimes even less honest? Are there not some whose vicinity extinguishes all faith and all enthusiasm, who draw you by your weaknesses, who govern you by your evil propensities, and make you die slowly to morality in a torment like that of Mezentius? These are dead people whom we mistake for living beings; these are vampires whom we regard as friends!

14 ב O

TRANSMUTATIONS

SPHERA LUNAE SEMPITERNUM AUXILIUM

ST. AUGUSTINE questioned seriously whether Apuleius could have been changed into an ass by a Thessalian sorceress, and theologians have long debated about the transformation of Nebuchadnezzar into a wild beast, which things merely prove that the eloquent doctor of Hippo was unacquainted with magical secrets and that the theologians in question were not advanced far in exegesis. We are concerned in this chapter with different and more incredible marvels, which are at the same time incontestable. I refer to lycanthropy, or the nocturnal transformation of men into wolves, long celebrated in country tales of the twilight by the histories of were-wolves. These histories are so well attested that, with a view to their explanation, skeptical science has recourse to furious mania and masquerading as animals. But such hypotheses are puerile and explain nothing. Let us turn elsewhere for the secret of the phenomena which have been observed on this subject and begin with establishing; 1, That no one has ever been killed by a were-wolf, except by suffocation, without effusion of blood and without wounds; 2, That were-wolves, though tracked, pursued and even wounded, have never been killed on the spot; 3, That persons suspected of these transformations have always been found at home, after a were-wolf chase, more or less maimed, sometimes dying, but invariably in their natural form.

Let us, next, establish phenomena of a different order. Nothing in the world is better borne out by evidence than the visible and real presence of P. Alphonsus Ligouri beside the dying pope, whilst the same personage was simultaneously seen at home, far from Rome, in prayer and ecstasy. Further, the simultaneous presence of the missionary Francis Xavier in several places at one time has been no less strictly demonstrated. It will be said that these are miracles, but we reply that miracles when they are genuine are simply facts for science. Apparitions of persons dear to us coincidently with the moment of their death are phenomena of the same order and attributable to the same cause. We have spoken of the sidereal body which is intermediary between the soul and the physical body. Now, this body frequently remains awake while the latter sleeps, and passes through all space which universal magnetism opens to it. It lengthens without breaking the sympathetic chain which attaches it to our heart and brain, and it is for this reason that it is so dangerous to awaken dreamers suddenly. As a fact, too great a start may break the chain in an instant and cause death immediately. The form of our sidereal body is conformed to the habitual condition of our thoughts, and it modifies, in the long run, the characteristics of the material body. That is why Swedenborg, in his somnambulistic intuitions, frequently beheld spirits in the shape of various animals.

Let us now make bold to say that a were-wolf is nothing else but the sidereal body of a man whose savage and sanguinary instincts are typified by the wolf; who, further, whilst his phantom wanders over the country, is sleeping painfully in his bed and dreams that he is actually a wolf. What makes the were-wolf visible is the almost somnambulistic excitement caused by the fright of those who behold it, or else the tendency, more particularly in simple country persons, to enter into direct communication with the astral light, which is the common medium of visions and dreams. The hurts inflicted on the were-wolf really wound the sleeping person by the odic and sympathetic congestion of the astral light, and by correspondence of the immaterial with the material body. Many persons will believe that they are dreaming when they read such things as these, and will ask whether we are really awake ourselves; but we will only request men of science to reflect upon the phenomena of gestation, and upon the influence of the imagination of women on the form of their offspring. A woman who had been present at the execution of a man who was broken on the wheel gave birth to a child with all its limbs shattered. Let anyone explain to us how the impression produced upon the soul of the mother by a horrible spectacle could so have reacted on the child, and we will explain in turn why blows received in dreams can really bruise and even grievously wound the body of him who receives them in imagination, above all when his body is suffering and subjected to nervous and magnetic influences.

To these phenomena and to the occult laws which govern them must be referred the effects of bewitchment, of which we shall speak hereafter. Diabolical obsessions, and the majority of nervous diseases which affect the brain, are wounds inflicted on the nervous mechanism by perverted astral light, meaning that which is absorbed or projected in abnormal proportions. All extraordinary and extra-natural tensions of the will predispose to obsessions and nervous diseases; enforced celibacy, asceticism, hatred, ambition, rejected love, are so many generative principles of infernal forms and influences. Paracelsus says that the menstruations of women beget phantoms in the air, and from this standpoint convents would be seminaries for nightmares, while the devils might be compared to those heads of the hydra of Lerne which were reproduced eternally and propagated in the very blood from their wounds. The phenomena of possession amongst the Ursulines of Loudun, so fatal to Urban Grandier, have been misconstrued. The nuns were really possessed by hysteria and fanatical imitation of the secret thoughts of their exorcists, which were transmitted to their nervous system by the astral light. They received the impression of all the hatreds which this unfortunate priest had conjured up against him, and such wholly interior communication seemed diabolical and miraculous to themselves. Hence in this tragical affair everyone acted sincerely, even to Laubardemont, who, in his blind execution of the prejudged verdicts of Cardinal Richelieu, believed that he was fulfilling at the same time the duties of a true judge, and as little suspected himself of being a follower of Pontius Pilate as he would have recognized in the skeptical and libertine curé of Saint-Pierre-du-Marché, a disciple and martyr of Christ. The possession of the nuns of Louvier is scarcely more than a copy of those of Loudun; the devils invent little and plagiarize one another. The process of Gaufridi and Magdalen de la Palud possesses stranger features, for in this case the victims are their own accusers. Gaufridi confessed that he was guilty of depriving a number of women of the power to defend themselves against his seductions by simply breathing in their nostrils. A young and beautiful girl, of noble family, who had been thus insufflated, described, in the greatest detail, scenes wherein the unchaste seemed to vie with the monstrous and grotesque. Such are the ordinary hallucinations of false mysticism and ill-kept celibacy. Gaufridi and his mistress were obsessed by their mutual chimeras, and the brain of the one reflected the nightmares of the other. Was not the Marquis of Sade himself infectious for certain depleted and diseased natures?

The scandalous trial of Father Girard is a new proof of the deliriums of mysticism and the singular nervous affections which it may entail. The trances of la Cadière, her ecstasies, her stigmata, were all as real as the insensate and perhaps involuntary debauchery of her director. She accused him, when he wished to withdraw from her, and the conversion of this young woman was a revenge, for there is nothing more cruel than depraved passions. An influential body, which intervened in the trial of Grandier for the destruction of the possible heretic, in this case rescued Father Girard for the honor of the order. Moreover, Grandier and Girard attained the same results by very different means, with which we shall be specially concerned in the sixteenth chapter.

We act by our imagination on the imagination of others, by our sidereal body on theirs, by our organs on their organs, in such a way that, by sympathy, whether of inclination or obsession, we reciprocally possess one another, and identify ourselves with those upon whom we wish to act. Reactions against such dominations frequently cause the most pronounced antipathy to succeed the keenest sympathy. Love has a tendency to unify beings; in thus identifying, it frequently renders them rivals, and, consequently, enemies, if in the depth of the two natures there is some unsociable disposition like pride. To permeate two united souls in an equal degree with pride is to disjoin them by making them rivals. Antagonism is the necessary consequence of a plurality of gods.

When we dream of a living person, either their sidereal body presents itself to ours in the astral light or at least a reflection thereof, and our impressions at the meeting often make known the secret dispositions of the person in our regard. For example, love fashions the sidereal body of the one in the image and likeness of the other, so that the psychal medium of the woman is like a man, and that of the man like a woman. It was this transfer which the Kabbalists sought to express in an occult manner when they said, in explanation of an obscure term in Genesis: "God created love by placing a rib of Adam in the breast of the woman, and a portion of the flesh of Eve in the breast of the man, so that at the bottom of woman's heart there is the bone of man, while at the bottom of man's heart there is the flesh of woman" – an allegory which is certainly not devoid of depth and beauty.

We have referred, in the previous chapter, to what the masters in Kabbalah call the embryonic condition of souls. This state, completed after the death of the person who thereby possesses another, is often commenced in life, whether by obsession or by love. I knew a young woman, whose parents inspired her with a great terror, and who took suddenly to inflict upon an inoffensive person the very acts she dreaded in them. I knew another who, after participating in an evocation concerned with a guilty woman suffering in the next world for certain eccentric acts, began to imitate, without any reason, the actions of the dead person. To this occult power must be attributed the terrible influence resident in parental malediction, which is feared by all nations on earth, as also the imminent danger of magical operations when anyone has not reached the isolation of true adepts. This virtue of sidereal transmutation, which really exists in love, explains the allegorical marvels of the wand of Circe. Apuleius speaks of a Thessalian woman who changed herself into a bird; he won the affections of her servant to discover the secrets of the mistress, but succeeded only in transforming himself into an ass. This allegory contains the most concealed secrets of love. Again, the Kabbalists say that when a man falls in love with a female elementary – undine, sylphide or gnomide, as the case may be – she becomes immortal with him, or otherwise he dies with her. We have already seen that elementaries are imperfect and as yet mortal men. The revelation we have mentioned, which has been regarded merely as a fable, is therefore the dogma of moral solidarity in love, which is itself the foundation of love, explaining all its sanctity and all its power. Who, then, is this Circe, that changes her worshippers into swine, while, so soon as she is subjected to the bond of love, her enchantments are destroyed? She is the ancient courtesan, the marble woman of all the ages. A woman who is without love absorbs and degrades all who come near her; she who loves, on the other hand, diffuses enthusiasm, nobility and life.

There was much talk in the last century about an adept accused of charlatanism, who was termed in his lifetime the divine Cagliostro. It is known that he practiced evocations, and that in this art he was surpassed only by the illuminated Schrœpffer.[17] It is said also that he boasted of his power in binding sympathies, and that he claimed to be in possession of the secret of the great work; but that which rendered him still more famous was a certain elixir of life, which immediately restored to the aged the strength and vitality of youth. The basis of this composition was malvoisie wine, and it was obtained by distilling the sperm of certain animals with the sap of certain plants. We are in possession of the recipe, but our reasons for withholding it will be

---

[17] See, in the *Ritual*, Schrœpffer's secrets and formulas for evocation.

readily understood.

15 ⬛ P

BLACK MAGIC

SAMAEL AUXILIATOR

We approach the mystery of black magic. We are about to confront, even in his own sanctuary, the black god of the Sabbath, the formidable goat of Mendes. At this point those who are subject to fear should close the book; even persons who are a prey to nervous impressions will do well to divert themselves or to abstain. We have set ourselves a task, and we must complete it. Let us first of all address ourselves frankly and boldly to the question: Is there a devil? What is the devil? As to the first point, science is silent, philosophy denies it on chance, religion only answers in the affirmative. As to the second point, religion states that the devil is the fallen angel; occult philosophy accepts and explains this definition. It will be unnecessary to repeat what we have already said on the subject; we will add here a further revelation:

> IN BLACK MAGIC, THE DEVIL IS THE GREAT MAGICAL AGENT EMPLOYED FOR EVIL PURPOSES BY A PERVERSE WILL.

The old serpent of the legend is nothing else than the universal agent, the eternal fire of terrestrial life, the soul of the earth, and the living fount of hell. We have said that the astral light is the receptacle of forms, and these when evoked by reason are produced harmoniously, but when evoked by madness they appear disorderly and monstrous; so originated the nightmares of St. Anthony and the phantoms of the Sabbath. Do, therefore, the evocations of goëtia and demonomania possess a practical result? Yes, certainly – one which cannot be contested, one more terrible than could be recounted by legends! When any one invokes the devil with intentional ceremonies, the devil comes, and is seen. To escape dying from horror at that sight, to escape catalepsy or idiocy, one must be already mad. Grandier was a libertine through indevotion[18], and perhaps also through skepticism; excessive zeal, following on the aberrations of asceticism and blindness of faith, depraved Girard, and made him deprave in his turn. In the fifteenth chapter of our Ritual we shall give all the diabolical evocations and practices of black magic, not that they may be used, but that they may be known and judged, and that such insanities may be put aside for ever.

M. Eudes de Mirville, whose book upon table-turning made a certain sensation recently, will possibly be contented and discontented at the same time with the solution here given of black magic and its problems. As a fact, we maintain, like himself, the reality and prodigious nature of the facts; with him also we assign them to the old serpent, the secret prince of this world; but we are not agreed as to the nature of this blind agent, which, under different directions, is the instrument of all good and of all evil, the minister of prophets and the inspirer of pythonesses. In a word, the devil, for us, is force placed temporarily at the disposal of error, even as mortal sin is, to our thinking, the persistence of the will in what is absurd. M. de Mirville is

---

[18] Spelling verified in the original edition. -pnw

therefore a thousand times right, but he is once and one great time wrong.

What we must exclude above all from the realm of existences is the arbitrary. Nothing happens by chance, nor yet by the autocracy of a good or evil will. There are two houses in heaven, and the lower house of Satan is restrained in its extremes by the senate of divine wisdom.

16 ל Q

BEWITCHMENTS

FONS OCULUS FULGUR

WHEN a man gazes unchastely upon any woman he profanes that woman, said the Great Master. What is willed with persistence is done. Every real will is confirmed by acts; every will confirmed by an act is action. Every action is subject to a judgement, and such judgement is eternal. These are dogmas and principles from which it follows that the good or evil which we will, to others as to ourselves, according to the capacity of our will and within the sphere of our operation, will infallibly take place, if the will be confirmed and the determination fixed by acts. The acts should be analogous to the will. The intent to do harm or to excite love, in order to be efficacious, must be confirmed by deeds of hatred or affection. Whatsoever bears the impression of a human soul belongs to that soul; whatsoever a man has appropriated after any manner becomes his body in the broader acceptation of the term, and anything which is done to the body of a man is felt, mediately or immediately, by his soul. It is for this reason that every species of hostility towards against one's neighbor is regarded in moral theology as the beginning of homicide. Bewitchment is a homicide, and the more infamous because it eludes self-defense by the victim and punishment by law. This principle being established to exonerate our conscience, and for the warning of weak vessels, let us affirm boldly that bewitchment is possible. Let us even go further and lay down that it is not only possible, but in some sense necessary and fatal. It is continually going on in the social world, unconsciously both to agents and patients. Involuntary bewitchment is one of the most terrible dangers of human life. Passional sympathy inevitably subjects the hottest desire to the strongest will. Moral maladies are more contagious than physical, and there are some triumphs of infatuation and fashion which are comparable to leprosy or cholera. We may die of an evil acquaintance as well as of a contagious touch, and the frightful plague which, during recent centuries only, has avenged in Europe the profanation of the mysteries of love, is a revelation of the analogical laws of nature, and at the same time offers only a feeble image of the moral corruptions which follow daily on an equivocal sympathy. There is a story of a jealous and infamous man who, to avenge himself on a rival, contracted an incurable disorder, and made it the common scourge and anathema of a divided bed. This atrocious history is that of every magician, or rather of every sorcerer who practices bewitchments. He poisons himself in order that he may poison others; he damns himself that he may torture others; he draws in hell with his breath in order that he may expel it by his breath; he wounds himself to death that he may inflict death on others; but possessed of this unhappy courage, it is positive and certain that he will poison and slay by the mere projection of his perverse will. There are some forms of love which are as deadly as hatred, and the bewitchments of goodwill are the torment of the wicked. The prayers offered to God for the conversion of a man bring misfortune to that man if he will not be converted. As we have already said, it is weariness and danger to strive against the fluidic currents occasioned by chains of wills in union.

Hence there are two kinds of bewitchment, voluntary and involuntary; physical and moral bewitchment may be also distinguished. Power attracts power, life attracts life, health attracts health; this is a law of nature. If two children live, above all, if they sleep together, and if one be weak while the other is strong, the strong will absorb the weak, and the latter will waste away. For this reason, it is important that children should always sleep alone. In conventual seminaries certain pupils absorb the intelligence of the others, and in every given circle of men, an individual speedily appears who avails himself of the wills of the rest. Bewitchment by means of currents is exceedingly common, as we have already observed; morally as well as physically, most of us are carried away by the crowd. What, however, we have proposed to exhibit more especially in this chapter is the almost absolute power of the human will upon the determination of its acts and the influence of every outward demonstration upon outward things.

Voluntary bewitchments are still frequent in our rural places because natural forces, among ignorant and isolated persons, operate without being diminished by any doubt or any diversion. A frank, absolute hatred, unleavened by rejected passion or personal cupidity is, under certain given conditions, a death-sentence for its object. I say unmixed with amorous passion or cupidity, because a desire, being an attraction, counterbalances and annuls the power of projection. For example, a jealous person will never efficaciously bewitch his rival, and a greedy heir will never by the mere fact of his will succeed in shortening the days of a miserly and long-lived uncle. Bewitchments attempted under such conditions reflect upon the operator and help rather than hurt their object, setting him free from a hostile action which destroys itself by excessive exaggeration. The term *envoûtement* (bewitchment) so strong in its Gaelic simplicity, expresses admirably what it means, the act of enveloping someone, so to speak, in a formulated will. The instrument of bewitchments is the great magic agent which, under the influence of an evil will, becomes really and positively the demon. Witchcraft, properly so called, that is, ceremonial operation with intent to bewitch, acts only on the operator, and serves to fix and confirm his will, by formulating it with persistence and labor, the two conditions which make volition efficacious. The more difficult or horrible the operation, the greater is its power, because it acts more strongly on the imagination and confirms effort in direct ratio of resistance. This explains the bizarre nature and even atrocious character of the operations in black magic, as practiced by the ancients and in the middle ages, the diabolical masses, administration of sacraments to reptiles, effusions of blood, human sacrifices, and other monstrosities, which are the very essence and reality of goëtia or nigromancy. Such are the practices which from all time have brought down upon sorcerers the just repression of the laws. Black Magic is really only a graduated combination of sacrileges and murders designed for the permanent perversion of a human will and for the realization in a living man of the hideous phantom of the demon. It is therefore, properly speaking, the religion of the devil, the cultus of darkness, hatred of good carried to the height of paroxysm: it is the incarnation of death and the persistent creation of hell.

The Kabalist Bodin, who has been erroneously considered of a feeble and superstitious mind, had no other motive in writing his *Demonomania* than that of warning people against dangerous incredulity. Initiated by the study of the Kabbalah into the true secrets of Magic, he trembled at the danger to which society was exposed by the abandonment of this power to the wickedness of men. Hence he attempted what at the present time M. Eudes de Mirville is attempting amongst ourselves; he gathered facts without interpreting them, and affirmed in the face of inattentive or pre-occupied science the existence of the occult influences and criminal operations of evil magic. In his own day Bodin attracted no more attention than will be given to M. Eudes de Mirville, because it is not enough to indicate phenomena and to prejudge their cause if we would influence earnest men; we must study, explain and demonstrate such cause, and this is precisely what we are ourselves attempting. Will better success crown our own efforts?

It is possible to die through the love of certain people as by their hate; there are absorbing passions, under the breath of which we feel ourselves depleted like the spouses of vampires. Not only do the wicked torment the good, but unconsciously the good torture the wicked. The gentleness of Abel was a long and painful bewitchment for the ferocity of Cain. Among evil men, the hatred of good originates in the very instinct of self-preservation; moreover, they deny that what torments them is good, and, for their own peace, are driven to deify and justify evil. In the sight of Cain, Abel was a hypocrite and coward, who abused the pride of humanity by his scandalous submissions to divinity. How much must this first murderer have endured before making such a frightful attack upon his brother? Had Abel understood, he would have been afraid. Antipathy is the presentiment of a possible bewitchment, either of love or hatred, for we find love frequently succeeding repulsion. The astral light warns us of coming influences by its action on the more or less sensible, more or less active, nervous system. Instantaneous sympathies, electric loves, are explosions of the astral light, which are as exactly and mathematically demonstrable as the discharge of strong magnetic batteries. Thereby we may see what unexpected dangers threaten an uninitiated person who is perpetually fooling with fire in the neighborhood of invisible powder-mines. We are saturated with the astral light, and we project it unceasingly to make room for and to attract fresh supplies. The nervous instruments, which are specially designed either for attraction or projection, are the eyes and hands. The polarity of the hand is resident in the thumb, and hence, according to the magical tradition which still lingers in rural places, whenever anyone is in suspicious company, he should keep the thumb doubled up and hidden in the hand, and while in the main avoiding a fixed glance at any one, still being the first to look at those whom we have reason to fear, so as to escape unexpected fluidic projections and fascinating regards.

There are certain animals which have the power of breaking the currents of astral light by an absorption peculiar to themselves. They are violently antipathetic to us, and possess a certain sorcery of the eye: the toad, the basilisk, and the tard are instances. These animals, when tamed and carried alive on the person, or kept in occupied rooms, are a guarantee against the hallucinations and trickeries of ASTRAL INTOXICATION, a term made use of here for the first time and one which explains all phenomena of unbridled passions, mental exaltations, and folly. Tame toads and tards, my dear sir, the disciple of Voltaire will say to me; carry them about with you, and write no more. To which I may answer, that I shall think seriously of doing so if ever I feel tempted to laugh at anything I do not understand, and to treat those whose knowledge and wisdom I fail to understand, as fools or as madmen. Paracelsus, the greatest of the Christian magi, opposed bewitchment by the practices of a contrary bewitchment. He composed sympathetic remedies and applied them, not to the suffering members, but to representations of these, formed and consecrated according to magical ceremonial. His successes were incredible, for never has any physician approached Paracelsus in his marvels of healing. But Paracelsus had discovered magnetism long before Mesmer, and had carried to its final consequences this luminous discovery, or rather this initiation into the magic of the ancients, who better than us understood the great magical agent and did not regard the astral light, azoth, the universal magnesia of the sages, as an animal and a special fluid emanating only from particular creatures. In his occult philosophy, Paracelsus opposes ceremonial magic, the terrible power of which he did not certainly ignore, but he sought to decry its practices so as to discredit black magic. He locates the omnipotence of the magus in the interior and occult *magnes*, and the most skillful magnetizers of our own day could not express themselves better. At the same time, he counselled the employment of magical symbols, talismans above all, in the cure of diseases. In our eighteenth chapter we shall have occasion to return to the talismans of Paracelsus, while following Gaffarel upon the great question of occult iconography and numismatics.

Bewitchment may be cured also by substitution, when that is possible, and by the rupture or deflection of the astral current. The rural traditions on all these points are admirable, and undoubtedly of remote antiquity; they are remnants of the instruction of the Druids, who were initiated in the mysteries of Egypt and India by wandering hierophants. Now, it is well known in vulgar magic that a bewitchment – that is, a will persistently confirmed in ill doing, invariably has its result, and cannot draw back without risk of death. The sorcerer who liberates any one from a charm must have another object for his malevolence, or it is certain that he himself will be smitten, and will perish as the victim of his own spells. The astral movement being circular, every azotic or magnetic emission which does not encounter its medium returns with force to its point of departure, thus explaining one of the strangest histories in a sacred book, that of the demons sent into the swine, which thereupon cast themselves into the sea. This act of high initiation was nothing else but the rupture of a magnetic current infected by evil wills. Our name is legion, for we are many, said the instinctive voice of the possessed sufferer. Possessions by the demon are bewitchments, and such cases are innumerable at the present day. A holy monk who has devoted himself to the service of the insane, Brother Hilarion Tissot, has succeeded, by long experience and incessant practice, in curing a number of patients by unconsciously using the magnetism of Paracelsus. He attributes most of his cases either to disorder of the will or to the perverse influence of external wills; he regards all crimes as acts of madness, and would treat the wicked as diseased, instead of exasperating and making them incurable, under the pretense of punishing them. What space of time must still elapse ere poor Brother Hilarion Tissot shall be hailed as a man of genius! And how many serious men, when they read this chapter, will say that Tissot and myself should treat one another according to our common ideas, but should refrain from publishing our theories, if we do not wish to be reckoned as physicians worthy of a hospital for incurables. It revolves, notwithstanding, said Galileo, stamping his foot upon the earth. Ye shall know the truth, and the truth shall make you free, said the Savior of men. It might also be added: Ye shall love justice, and justice shall make you whole men. A vice is a poison, even for the body; true virtue is a pledge of longevity.

The method of *ceremonial bewitchments* varies with times and persons;

all subtle and domineering people find its secrets and its practice within themselves, without even actually calculating about them or reasoning on their sequence. Herein they follow instinctive inspirations of the great agent, which, as we have already said, accommodates itself marvelously to our vices and our virtues; it may, however, be generally laid down that we are subjected to the wills of others according to the analogies of our tendencies, and above all, of our faults. To pamper the weaknesses of an individuality is to possess ourselves of that individuality and convert it into an instrument in the order of the same errors or depravities. Now, when two natures whose defects are analogous become subordinated one to another, the result is a sort of substitution of the stronger for the weaker, an actual obsession of one mind by the other. Very often the weaker may struggle and seek to revolt, but it falls only deeper in servitude. So, did Louis XIII conspire against Richelieu and subsequently, so to speak, sought his pardon by abandoning his accomplices. We have all a ruling defect, which is for our soul as the umbilical cord of its sinful birth, and it is by this that the enemy can always seize us – for some vanity, for others idleness, for the majority egotism. Let a wicked and crafty mind avail itself of this snare and we are lost; we may not go mad or turn idiots, but we become positively alienated, in all the force of the expression – that is, we are subjected to a foreign impulsion. In such a state one dreads instinctively everything that might bring us back to reason, and will not even listen to representations that are opposed to our infatuation. Here is one of the most dangerous disorders which can affect the moral nature. The sole remedy for such a bewitchment is to make use of madness itself in order to cure madness, to provide the sufferer with imaginary satisfactions in the opposite order to that wherein he is now immersed. Endeavour, for example, to cure an ambitious person by making him desire the glories of heaven – mystic remedy; cure one who is dissolute by true love – natural remedy; obtain honorable successes for a vain person; exhibit unselfishness to the avaricious, and procure for them legitimate profit by honorable participation in generous enterprises, etc. Acting in this way upon the moral nature, we may succeed in curing a number of physical maladies, for the moral affects the physical in virtue of the magical axiom: "That which is above is like unto that which is below." This is why the Master said, when speaking of the paralyzed woman: Satan has bound her. A disease invariably originates in a deficiency or an excess, and ever at the root of a physical evil we shall find a moral disorder. This is an unchanging law of nature.

17  R

ASTROLOGY

STELLA OS INFLEXUS

OF all the arts which have originated in ancient magian[19] wisdom astrology is now the most misunderstood. No one believes any longer in the universal harmonies of nature and in the necessary interlacing of all effects with all causes. Moreover, true astrology, that which connects with the unique and universal dogma of the Kabbalah, became profaned among the Greeks and Romans of the decline. The doctrine of the seven spheres and the three mobiles, drawn primitively from the sephirotic decade, the character of the planets governed by angels, whose names have been changed into those of Pagan divinities, the influence of the spheres on one another, the destiny attached to numbers, the scale of proportion between the celestial hierarchies corresponding to the human hierarchies – all this has been materialized and degraded into superstition by genethliacal[20] soothsayers and erectors of horoscopes during the decline and the middle ages. The restoration of astrology to its primitive purity would be, in a sense, the creation of an entirely new science; here let us

---

[19] Spelling as in the original edition. -pnw
[20] Ibid

attempt merely to indicate its first principles, with their more immediate and approximate consequences.

We have said that the astral light receives and preserves the impressions of all visible things; it follows from this that the daily position of the heaven is reflected in this light, which, being the chief agent of life, operates the conception, gestation and birth of children by a sequence of apparatuses naturally designed to this end. Now, if this light be sufficiently prodigal of images as to impart the fruit of the womb the visible imprints of a maternal fantasy or appetite, still more will it transmit to the plastic and indeterminate temperament of a newly-born child the atmospheric impressions and diverse influences which, in the entire planetary system, are consequent at a given moment upon such or such particular aspect of the stars. Nothing is indifferent in nature; a stone more or a stone less upon a road may break or modify profoundly the destinies of the greatest men or even the largest empires; still more must the position of this or that star in the sky have an influence on the child who is born, who enters by the very fact of his birth into the universal harmony of the sidereal world. The stars are bound to one another by the attractions which hold them in equilibrium and cause them to move with uniformity through space. From all spheres unto all spheres there stretch these indestructible threads of light, and there is no point upon any planet to which one of them is not attached. The true adept in astrology must, therefore, give heed to the precise time and place of the birth which is in question; then, after an exact calculation of the astral influences, it remains for him to compute the chances of estate, that is to say, the advantages or hindrances which the child must one day meet with by reason of position, relatives, inherited tendencies, and hence natural proclivities, in the fulfilment of his destinies. Finally, he will still have to take into consideration human liberty and its initiative, should the child eventually come to be a true man, and to isolate himself by an intrepid will from fatal influences and from the chain of destiny. It will be seen that we do not allow too much to astrology, but so much as we leave it is indubitable; it is the scientific and magical calculus of probabilities.

Astrology is as ancient as astronomy, and indeed it is more ancient; all seers of lucid antiquity have accorded it their fullest confidence; now, we must not condemn and reject in a shallow manner anything which comes before us protected and supported by such imposing authorities. Long and patient observations, conclusive comparisons, frequently repeated experiments, must have led the old sages to their decisions, and to refute them the same labor must be undertaken from an opposite standpoint. Paracelsus was perhaps the last of the great practical astrologers; he cured diseases by talismans formed under astral influences; he distinguished upon all bodies the mark of their dominant star; there, according to him, was the true universal medicine, the absolute science of nature, lost by man's own fault and recovered only by a small number of initiates. To recognize the sign of each star upon men, animals, and plants, is the true natural science of Solomon, that science which is said to be lost, but the principles of which are preserved notwithstanding, as are all other secrets, in the symbolism of the Kabbalah. It will be readily understood that in order to read the stars one must know the stars themselves; now, this knowledge is obtained by the kabbalistic *domification* of the sky and by the understanding of the celestial planisphere, recovered and explained by Gaffarel. In this planisphere the constellations form Hebrew letters, and the mythological figures may be replaced by the symbols of the Tarot. To this same planisphere Gaffarel refers the origin of patriarchal writing, and the first lineaments of primitive characters may very well have been found, in which case the celestial book would have served as the model of Henoch's, and the kabbalistic alphabet would have been a synopsis of the entire sky. This is not wanting in poetry, nor, above all, in probability, and the study of the Tarot, which is evidently the primitive and hieroglyphic work of Henoch, as was divined by the erudite William Postel, is sufficient to convince us hereof.

The signs imprinted in the astral light by the reflection and attraction of the stars are reproduced, therefore, as the sages have discovered, on all bodies which are formed by the co-operation of that light. Men bear the signs of their star on their forehead chiefly, and in their hands; animals in their whole form, and in their individual signs; plants in their leaves and seed; minerals in their veins and their grain. The study of these characters was the entire life-work of Paracelsus, and the figures on his talismans are the result of his researches; he has, however, left us no key to them, so that the astral kabbalistic alphabet with its correspondences still remains to be done; as regards publicity, the science of nonconventional magical writing stopped with the planisphere of Gaffarel. The serious art of divination rests wholly in the knowledge of these signs. Chiromancy is the art of reading the writing of the stars in the lines of the hand, and physiognomy seeks the same or analogous characters upon the countenance of its inquirers. As a fact, the lines formed on the human face by nervous contractions are determined fatally, and the radiation of the nervous tissue is absolutely analogous to those networks which are formed between the worlds by the chains of starry attraction. The fatalities of life are, therefore, written necessarily in our wrinkles, and a first glance frequently reveals upon the forehead of a stranger either one or more of the mysterious letters of the kabbalistic planisphere. Should the letter be jagged and laboriously inscribed, there has been a struggle between will and fatality, and in his most powerful emotions and tendencies, the individual's entire past manifests to the magus; from this it becomes easy to conjecture the future, and if events occasionally deceive the sagacity of the diviner, he who has consulted him will remain none the less astounded and convinced by the superhuman knowledge of the adept.

The human head is formed upon the model of the celestial spheres; it attracts and it radiates, and in the conception of a child, this it is which first forms and manifests. Hence the head is subject in an absolute manner to astral influence, and evidences its several attractions by its diverse protuberances. The final word of phrenology is to be found, therefore, in scientific and purified astrology, the problems of which we point out to the patience and good faith of scholars.

According to Ptolemy, the sun dries up and the moon moistens; according to the Kabalists, the sun represents rigorous Justice, while the moon is in sympathy with Mercy. It is the sun which produces storms, and, by a kind of gentle atmospheric pressure, the moon occasions the ebb and flow, or, as it were, the respiration of the sea. We read in the Zohar, one of the great sacred books of the Kabbalah, that "the magical serpent, the son of the Sun, was about to devour the world, when the Sea, daughter of the Moon, set her foot upon his head and subdued him." For this reason, among the ancients, Venus was the daughter of the Sea, as Diana was identical with the Moon. Hence also the name of Mary signifies star or salt of the sea. To consecrate this kabbalistic doctrine in the belief of the vulgar, it is said in prophetic language: The woman shall crush the serpent's head.

Jerome Cardan, one of the boldest students, and beyond contradiction the most skillful astrologer of his time – Jerome Cardan, who, if we accept the legend of his death, was a martyr to his faith in astrology, has left behind him a calculation by means of which anyone can foresee the good or evil fortune special to all the years of his life. His theory was based upon his own experiences, and he assures us that the calculation never deceived him. To ascertain the fortune of a given year, he sums up the events of those which have preceded it by 4, 8, 12, 19 and 30; the number 4 is that of realization; 8 is the number of Venus or natural things; 12 belongs to the cycle of Jupiter, and corresponds to successes; 19 has reference to the cycles of the Moon and of Mars; the number 30 is that of Saturn or Fatality. Thus, for example, I desire to ascertain what will befall me in this present year, 1855. I pass therefore in review those decisive events in the order of life and progress which occurred four years ago; the natural felicity or misfortune of eight years back; the successes or failures belonging to twelve years since; the vicissitudes and miseries or diseases which overtook me nineteen years from now, and my tragic or fatal experiences of thirty years back. Then, taking into account irrevocably accomplished facts and the advance of time, I calculate the chances analogous to those which I owe already to the influence of the same planets, and I conclude that in 1851 I had employment which was moderately but sufficiently remunerative, with some embarrassment of position; in 1847 I was separated violently from my family, with great attendant sufferings for mine and me; in 1843 I travelled as an apostle, addressing the people, and suffering the persecution of ill-meaning persons: briefly, I was at once honored and proscribed. Finally, in 1825 family life came to an end for me, and I entered definitely on that fatal path which led me to science and misfortune. I may suppose therefore that this year I shall experience toil, poverty, vexation, heart-exile, change of place, publicity and contradictions, with some eventuality which will be decisive for the rest of my life: every indication in the present leads me to endorse this forecast. Hence I conclude that, for myself and for this year, experience confirms fully the precision of Cardan's astrological calculus, which, furthermore, connects with the climacteric years of ancient astrologers. This term signifies arranged in scales or calculated on the degrees of a scale. Johannes Trithemius in his book on *Secondary Causes* has very curiously computed the return of fortunate or calamitous years for all the empires of the world. In the twenty-first chapter of our Ritual we shall give an exact analysis of this work, together with a continuation of the labor of Trithemius to our own days and the application of his magical scale to contemporary events, so as to deduce the most striking probabilities relative to the immediate future of France, Europe and the world.

According to all the grand masters in astrology, comets are the stars of exceptional heroes, and they visit earth only to signalize great changes; the planets preside over collective existences and modify the destinies of mankind in the aggregate; the fixed stars, more remote and more feeble in their action, attract individuals and determine their tendencies; sometimes a group of stars combine to influence the destinies of a single man, while often a great number of souls are driven by the distant rays of the same sun. When we die, our interior light in departing follows the attraction of its star, and thus it is that we live in other universes, where the soul makes for itself a new garment, analogous to the development or diminution of its beauty; for our souls, when separated from our bodies, resemble revolving stars; they are globules of animated light which always seek their center for the recovery of their equilibrium and their true movement. Before all things, however, they must liberate themselves from the folds of the serpent, that is, the unpurified astral light which envelops and imprisons them, unless the strength of their will can lift them beyond its reach. The immersion of the living star in the dead light is a frightful torment, comparable to that of Mezentius. Therein the soul freezes and burns at the same time and has no means of getting free except by re-entering the current of exterior forms and assuming a fleshly envelope, then energetically battling against instincts to strengthen that moral liberty which will permit it at the moment of its death to break the chains of earth and wing its flight in triumph towards the star of consolation which has smiled in light upon it. Following this clue, we can understand the nature of the fire of hell, which is identical with the demon or old serpent; we can gather also wherein consist the salvation and reprobation of men, all called and all successively elected, but in small number, after having risked falling into the eternal fire through their own fault.

Such is the great and sublime revelation of the magi, a revelation which is the mother of all symbols, of all dogmas, of all religions. We can realize already how far Dupuis was mistaken in regarding astronomy as the source of every cultus. It is astronomy, on the contrary, which has sprung from astrology, and primitive astrology is one of the branches of the holy Kabbalah, the science of sciences, and the religion of religions. Hence upon the seventeenth page of the Tarot we find an admirable allegory – a naked woman, typifying Truth, Nature and Wisdom at one and the same time, turns two ewers towards earth, and pours out fire and water upon it; above her head glitters the septenary, starred about an eight-pointed star, that of Venus, symbol of peace and love; the plants of earth are flourishing around the woman, and on one of them the butterfly of Psyche has alighted; this emblem of the soul is replaced in some copies of the sacred book by a bird, which is a more Egyptian and probably a more ancient symbol. In the modern Tarot the plate is entitled the Glittering Star; it is analogous to a number of Hermetic symbols, and is also in correspondence with the Blazing Star of Masonic initiates, which expresses most of the mysteries of Rosicrucian secret doctrine.

18 א S

CHARMS AND PHILTRES

JUSTITIA MYSTERIUM CANES

WE have now to grapple with the most criminal abuse to which magical sciences can be put, namely, venomous magic, or, rather, sorcery. Let it be understood here that we write not to instruct but to warn. If human justice, instead of punishing the adepts, had only proscribed the necromancers and poisoning sorcerers, it is certain, as we have previously remarked, that its severity would have been well placed, and that the most severe penalties could never be excessive in the case of such criminals. At the same time it must not be supposed that the right of life and death which secretly belongs to the magus has always been exercised to satisfy some infamous vengeance, or some cupidity more infamous still. In the middle ages, as in the ancient world, magical associations have frequently struck down or destroyed slowly the revealers or profaners of mysteries, and when the magic sword has refrained from striking, when the spilling of blood was dangerous, then Aqua Toffana, poisoned nosegays, the shirt of Nessus, and other deadly instruments, still stranger and still less known, were used to carry out sooner or later the terrible sentence of the free judges. We have said that there is in magic a great and indicible[21] arcanum, which is never mentioned among adepts, which the profane above all must be prevented from divining; in former times, whosoever revealed, or caused the key of this supreme secret to be discovered by others through imprudent revelations, was condemned immediately to death, and was often driven to execute the sentence himself. The celebrated prophetic supper of Cazotte, described by Laharpe, has not been understood hitherto. Laharpe very naturally yielded to the temptation of surprising his readers by amplifying the details of his narrative. Everyone present at this supper, Laharpe excepted, was an initiate and a divulger, or at least a profaner, of the mysteries. Cazotte, the most exalted of all in the scale of initiation, pronounced their sentence of death in the name of illuminism, and this sentence was variously but rigorously executed, even as several years and several centuries previously had occurred in the case of similar judgements against the Abbé de Villars, Urban Grandier, and many others. The revolutionary philosophers perished as did Cagliostro deserted in the prisons of the Inquisition, as did the mystic band of Catherine Theos, as did the imprudent Schrœpffer, constrained to suicide in the midst of his magical triumphs and the universal infatuation, as did the deserter Kotzebuë, who was stabbed by Carl Sand, as did also so many others whose corpses have been discovered without any one being able to learn the cause of their sudden and sanguinary death. The strange allocution addressed to Cazotte when he himself was condemned by the president of the revolutionary tribunal will be called readily to mind. The Gordian Knot of the terrible drama of '93 is still concealed in the darkest sanctuary of the secret societies; to adepts of good faith, who sought to emancipate the common people, were opposed adepts of another sect, attached to more ancient traditions, who fought by

---

[21] Spelling as in the original printed edition. -pnw

means analogous to those of their adversaries: the practice of the great arcanum was made impossible by unmasking its theory. The crowd understood nothing, but it mistrusted everything, and fell lower still in its discouragement; the great arcanum became more secret than ever; the adepts, checkmated by each other, could exercise their power neither to govern others nor to deliver themselves; they condemned one another to the death of traitors; they abandoned one another to exile, to suicide, to the knife and the scaffold.

I shall be asked possibly whether equally terrible dangers threaten at this day the intruders into the occult sanctuary and the betrayers of its secret. Why should I answer anything to the incredulity of the inquisitive? If I risk a violent death for their instruction, certainly they will not save me; if they are afraid on their own account, let them abstain from imprudent research – this is all I can say to them. Let us return to venomous magic.

In his romance of *Monte Christo*, Alexandre Dumas has revealed some practices of this ominous science. There is no need to traverse the same ground by repeating its melancholy theories of crime; describing how plants are poisoned; how animals nourished on these plants have their flesh infected, and becoming in turn the food of men, cause death without leaving any trace of poison; how the walls of houses are inoculated; how the air is permeated by fumes which require the glass mask of St. Croix for the operator; let us leave the ancient Canidia her abominable mysteries, and refrain from investigating the extent to which the infernal rites of Sagana have carried the art of Locusta. It is enough to state that this most infamous class of malefactors distilled in conjunction the virus of contagious diseases, the venom of reptiles, and the sap of poisonous plants, that they extracted from the fungus its deadly and narcotic properties, its asphyxiating principles from *datura stramonium*, from the peach and bitter almond that poison one drop of which, placed on the tongue or in the ear, destroys, like a flash of lightning, the strongest and best constituted living being. The white juice of sea-lettuce was boiled with milk in which vipers and asps had been drowned. The sap of the manchineel or deadly fruit of Java was either brought back with them from their long journeys, or imported at great expense; so also was the juice of the cassada, and so were similar poisons; they pulverized flint, mixed with impure ashes the dried slime of reptiles, composed hideous philtres with the virus of mares on heat and similar secretions of bitches; they mingled human blood with infamous drugs, composing an oil the mere odor of which was fatal, therein recalling the *tarte bourbonnaise* of Panurge; they even concealed recipes for poisoning in the technical language of alchemy, and the secret of the powder of projection, in more than one old book which claims to be Hermetic, is in reality that of the powder of succession. The *Grand Grimoire* gives one in particular which is very thinly disguised under the title of *Method of Making Gold*; it is an atrocious decoction of verdigris, arsenic and sawdust, which, if properly made, should immediately consume a branch that is plunged into it and eat swiftly through an iron nail. John Baptista Porta cites in his *Natural Magic* a specimen of Borgia poison, but, as may be imagined, he is deceiving the vulgar, and does not divulge the truth, which would be too dangerous in such a connection. We may therefore quote his recipe to satisfy the curiosity of our readers.

The toad itself is not venomous, but it is a sponge for poisons, and is the mushroom of the animal kingdom. Take, then, a plump toad, says Porta, and place it with vipers and asps in a globular bottle; let poisonous fungi, fox-gloves, and hemlock be their sole nourishment during a period of several days; then enrage them by beating, burning and tormenting in every conceivable manner, till they die of rage and hunger; sprinkle their bodies with powdered spurge and ground glass; then place them in a well-sealed retort, and extract all their moisture by fire. Let the glass cool; separate the ash of the dead bodies from the incombustible dust, which will remain at the bottom of the retort. You will then have two poisons – one liquid, the other a powder. The first will be fully as efficacious as the terrible *Aqua Poffana*; the second, in a few days' time, will cause any person, who may have a pinch of it mixed with his drink, to become, in the first place, wilted and old, and subsequently to die amidst horrible sufferings, or in a state of complete collapse. It must be admitted that this recipe has a magical physiognomy of the blackest and most revolting kind, and sickens one by its recollections of the abominable confections of Canidia and Medea. The sorcerers of the middle ages pretended to receive such powders at the Sabbath and sold them at a high price to the malicious and ignorant. The tradition of similar mysteries spread terror in country places and came to act as a spell. The imagination once impressed, the nervous system once assailed, the victim rapidly wasted away, the very dread of his relatives and friends sealing his loss. The sorcerer and sorceress were almost invariably a species of human toad, swollen with long-enduring rancours. They were poor, repulsed by all, and consequently full of hatred. The fear which they inspired was their consolation and their revenge; poisoned themselves by a society of which they had experienced nothing but the refuse and the vices, they poisoned in their turn all those who were weak enough to fear them, and avenged upon beauty and youth their accursed old age and their atrocious ugliness. The mere operation of these evil works, and the fulfilment of these loathsome mysteries constituted and confirmed what was then called a compact with the devil. It is certain that the worker must have been given over body and soul to evil, and justly deserved the universal and irrevocable reprobation expressed by the allegory of hell. That human souls could descend to such an abyss of crime and madness must assuredly astonish and grieve us; but is not such an abyss needed as a basis for the exaltation of the most sublime virtues? Does not the depth of infernus demonstrate by antithesis the infinite height and grandeur of heaven?

In the North, where the instincts are more repressed and vivacious; in Italy, where the passions are more diffusive and fiery, charms and the evil eye are still dreaded; the *jettatura* is not to be braved with impunity in Naples, and persons who are unfortunately endowed with this power are even distinguished by certain exterior signs. In order to guard against it, experts affirm that horns must be carried on the person, and the common people, who take everything literally, hasten to adorn themselves with small horns, not dreaming of the sense of the allegory. These attributes of Jupiter Ammon, Bacchus, and Moses are the symbol of moral power or enthusiasm, so that the magicians mean to say that, in order to withstand the jettatura, the fatal current of instincts must be governed by great intrepidity, a great enthusiasm, or a great thought. In like manner, almost all popular superstitions are profane interpretations of some grand maxim or marvelous secret of occult wisdom. Did not Pythagoras, in his admirable symbols, bequeath a perfect philosophy to sages, and a new series of vain observances and ridiculous practices to the vulgar? Thus, when he said: "Do not pick up what falls from the table; do not cut down trees on the great highway; kill not the serpent when it slips into your garden," – was he not inculcating the precepts of charity, either social or personal, under transparent allegories? When he said: "Do not look at yourself by torchlight in a mirror," was he not ingeniously teaching that true self-knowledge which is incompatible with factitious lights and the prejudgments of systems? It is the same with the other precepts of Pythagoras, who, it is well known, was followed literally by a swarm of unintelligent disciples, and, indeed, amongst our provincial superstitious observances, there are many which indubitably belong to the primitive misconception of Pythagorean symbols.

Superstition is derived from a Latin word which signifies survival. It is the sign surviving the thought; it is the dead body of a religious rite. Superstition is to initiation what the notion of the devil is to that of God. This is the sense in which the worship of images is forbidden, and in this sense also a doctrine most holy in its original conception may become superstitious and impious when it has lost its spirit and its inspiration. Then does religion, ever one, like the supreme reason, change its vestures and abandon old rites to the cupidity and roguery of priests dispossessed, and metamorphosed by their wickedness and ignorance into jugglers and charlatans. We may include among superstitions those magical emblems and characters, of which the meaning is no longer understood, which are engraved by chance on amulets and talismans. The magical images of the ancients were pantacles, *i.e.*, kabbalistic syntheses. Thus, the wheel of Pythagoras is a pantacle analogous to the wheels of Ezekiel; the two emblems contain the same secrets, and belong to the same philosophy; they constitute the key of all pantacles, and we have already discoursed concerning them.

The four beasts, or rather, the four-headed sphinx of the same prophet are identical with the admirable Indian symbol which we have reproduced in this work, as having reference to the great arcanum. In his Apocalypse, St. John followed and elaborated Ezekiel; indeed, the monstrous figures of his wonderful book are so many magical pantacles, the key of which is easily discoverable by kabbalists. On the other hand, Christians, rejecting science in their anxiety to extend faith, sought later on to conceal the origin of their dogmas, and condemned all kabbalistic and magical books to the flames. To destroy originals gives a kind of originality to copies, as was doubtless in the mind of St. Paul when, prompted beyond question by the most laudable intention, he accomplished his scientific *auto-da-fé* at Ephesus. In the same way, six centuries later, the true believer Omar sacrificed the Library of Alexandria to the originality of the Koran, and who knows whether in the time to come a future Apostle will not set fire to our literary museums and confiscate the printing-press in the interest of some fresh religious infatuation, some newly accredited legend?

The study of talismans and pantacles is one of the most curious branches of magic, and connects with historical numismatics. There are Indian, Egyptian, and Greek talismans, kabbalistic medals coming from ancient and modern Jews, Gnostic abraxas, occult tokens in use among the members of secret societies, and sometimes called counters of the Sabbath; so also, there are Templar medals and jewels of Freemasonry. In his *Treatise on the Wonders of Nature*, Coglenius describes the talismans of Solomon and those of Rabbi Chaël. Designs of many others that are most ancient will be found in the magical calendars of Tycho Brahé and Duchentau, and should have a place in M. Ragon's archives on initiation, a vast and scholarly undertaking, to which we refer our readers.

19 ק T

THE STONE OF THE PHILOSOPHERS – ELAGABALUS
VOCATIO SOL AURUM

THE ancients adored the Sun under the figure of a black stone, which they named Elagabalus, or Heliogabalus. What did this stone signify, and how came it to be the image of the most brilliant of luminaries? The disciples of Hermes, before promising their adepts the elixir of long life, or the powder of projection, counselled them to seek for the philosophical *stone*. What is this *stone*, and why a stone? The great initiator of the Christians invites his believers to build on the stone, or rock, if they do not wish their structures to be demolished. He terms Himself the corner-stone, and says to the most faithful of his Apostles, "Thou art Peter (*petrus*), and upon this rock (*petram*) I will build my church." This *stone*, say the masters in alchemy, is the true salt of the philosophers, which is the third ingredient in the composition of Azoth. Now, we know already that AZOTH is the name of the great Hermetic and true philosophical agent; furthermore, their Salt is represented under the figure of a cubic stone, as may be seen in the *Twelve Keys* of Basil Valentine,[22] or in the allegories of Trevisan. Once

---

[22] Volume 1 in the R.A.M.S. Library of Alchemy. -pnw

more, what is this stone actually? It is the foundation of absolute philosophy, it is supreme and immovable reason. Before even dreaming of the metallic work, we must be fixed for ever upon the absolute principles of wisdom, we must possess that reason which is the touch-stone of truth. Never will a man of prejudices become the king of nature and the master of transmutations. The philosophical stone is hence before all things necessary; but how is it to be found? Hermes informs us in his Emerald Table.[23] We must separate the subtle from the fixed with great care and assiduous attention. Thus, we must separate our certitudes from our beliefs, and sharply distinguish the respective domains of science and faith, understanding thoroughly that we do not know things which we believe, and that we cease immediately to believe anything which we come actually to know, so that the essence of the things of faith is the unknown and the indefinite, while it is quite the reverse with the things of science. It must thence be inferred that science rests on reason and experience, whilst the basis of faith is sentiment and reason. In other words, the philosophical stone is the true certitude which human prudence assures to conscientious researches and modest doubt, whilst religious enthusiasm ascribes it exclusively to faith. Now, it belongs neither to reason without aspirations nor to aspirations without reason; true certitude is the reciprocal acquiescence of the reason which knows in the sentiment which believes and of the sentiment which believes in the reason which knows. The permanent alliance of reason and faith will result not from their absolute distinction and separation, but from their mutual control arid their fraternal concurrence. Such is the significance of the two pillars of Solomon's porch, one named Jakin and the other Boaz, one black and the other white. They are distinct

---

[23] Volume 4 in the R.A.M.S. Library of Alchemy. -pnw

and separate, they are even contrary in appearance, but if blind force sought to join them by bringing them close to one another, the roof of the temple would collapse; separately, their power is one; joined, they are two powers which destroy one another. For precisely the same reason the spiritual power is weakened whenever it attempts to usurp the temporal, while the temporal power becomes the victim of its encroachments on the spiritual. Gregory VII ruined the Papacy; the schismatic kings have lost and will lose the monarchy. Human equilibrium requires two feet, the worlds gravitate by means of two forces, generation needs two sexes. Such is the meaning of the arcanum of Solomon, represented by the two pillars of the temple, Jakin and Boaz.

The sun and moon of the alchemists correspond to the same symbol and concur in the perfection and stability of the philosophical stone. The sun is the hieroglyphic sign of truth, because it is the visible source of light, and the rude stone is the symbol of stability. It was for this reason that the ancients took the stone Elagabalus as the actual type of the sun, and for this reason the mediaeval alchemists pointed to the philosophical stone as the first means of making philosophical gold, that is to say, of transforming the vital forces represented by the six metals into Sol, that is, into truth and light, the first and indispensable operation of the great work, leading to the secondary adaptations, and discovering, by the analogies of nature, the natural and grosser gold to the possessors of the spiritual and living gold, of the true salt, the true mercury and the true sulphur of the philosophers. To find the philosophical stone is then to have discovered the absolute, as the masters say otherwise. Now, the absolute is that which admits of no errors, it is the fixation of the volatile, it is the rule of the imagination, it is the very necessity of being, it is the immutable law of reason and truth; the absolute is that which is. Now that which is in some sense precedes he who is. God himself cannot be in the absence of a reason of being, and can exist only in virtue of a supreme and inevitable reason. It is this reason which is the absolute; it is this in which we must believe if we desire a rational and solid foundation for our faith. It may be said in these days that God is merely a hypothesis, but the absolute reason is not; it is essential to being.

St. Thomas once said: "A thing is not just because God wills it, but God wills it because it is just." Had St. Thomas logically deduced all the consequences of this beautiful thought, he would have found the philosophical stone, and besides being the angel of the school, he would have been their reformer. To believe in the reason of God and in the God of reason is to render atheism impossible. When Voltaire said: "If God did not exist, it would be necessary to invent Him," he felt rather than understood the reason which is in God. Does God really exist? There is no knowing, but we desire it to be so, and hence we believe it. Faith thus formulated is reasonable faith, for it admits the doubt of science, and, as a fact, we believe only in things which seem to us probable, though we do not know them. To think otherwise is delirium; to speak otherwise is to talk like illuminated or fanatical. Now, it is not to such persons that the philosophical stone is promised. The ignoramuses who have turned primitive Christianity from its path by substituting faith for science, dream for experience, the fantastic for the real; inquisitors who, during so many ages, have waged a war of extermination against magic; have succeeded in enveloping with darkness the ancient discoveries of the human mind, so that we are now groping for a key to the phenomena of nature. Now, all-natural phenomena depend upon a single and immutable law, represented by the philosophical stone, and especially by its cubic form. This law, expressed by the tetrad in the Kabbalah, equipped the Hebrews with all the mysteries of their divine Tetragram. It may be said therefore that the Philosophical Stone is square in every sense, like the heavenly Jerusalem of St. John; that one of its sides is inscribed with the name שלמה and the other with that of GOD; that one of its facets bears the name of ADAM, a second that of HEVA, and the two others those of AZOT and INRI. At the beginning of the French translation of a book by the Sieur de Nuisement on the philosophical salt, the spirit of the earth is represented standing on a cube over which tongues of flame are passing; the phallus is replaced by a caduceus; the sun and moon figure on the right and left breast; the is bearded, crowned and holds a scepter in his hand. This is the *Azoth* of the sages on its pedestal of salt and sulphur. The symbolic head of the goat of Mendes is occasionally given to this figure, and it is then the Baphomet of the Templars and the Word of the Gnostics – bizarre images which became scarecrows for the vulgar after affording food for thought to the sages, innocent hieroglyphs of thought and faith which have been a pretext for the rage of persecutions. How pitiable are men in their ignorance, but how they would despise themselves if only they came to know!

20 ד U

THE UNIVERSAL MEDICINE
CAPUT RESURRECTIO CIRCULUS

THE majority of our physical complaints come from our moral diseases, according to the one and universal dogma, and by reason of the law of analogies. A great passion to which we abandon ourselves corresponds always to a great malady in store for us. Mortal sins are so named because they, physically and positively, cause death. Alexander the Great died of pride; he was naturally temperate, and it was through pride that he yielded to the excess which occasioned his death. Francis I died of an adultery. Louis XV died of his deer-park. When Marat was assassinated he was perishing of rage and envy. He was a monomaniac of pride, who believed himself to be the only just man, and would have slain everything that was not Marat. Several of our contemporaries died of fallen ambition after the revolution of February. So soon as any will is confirmed irrevocably in a tendency towards the absurd, the man is dead, and the rock on which he will break is not distant.

It is, therefore, true to say that wisdom preserves and prolongs life.

The great Master told us: "My flesh is meat indeed, and My blood is drink indeed. He that eateth My flesh and drinketh My blood hath everlasting life." And when the crowd murmured, He added: "Here the flesh profiteth nothing: the words that I speak unto you are spirit and life." So also, when He was about to die, He attached the remembrance of His life to the sign of bread, and that of His spirit to the symbol of wine, thus instituting the communion of faith, hope and charity. Now, it is in the same sense that the Hermetic masters have said: Make gold potable, and you will have the Universal Medicine – that is to say, appropriate truth to your needs, let it become the source at which you daily drink, and you will in yourself have the immortality of the sages. Temperance, tranquility of soul, simplicity of character, calmness and rationality of will, these things not only make man happy, but strong and well-seeming. By growth in reason and goodness man becomes immortal. We are the authors of our own destiny, and God does not save us apart from our own concurrence. There is no death for the sage; death is a phantom, made horrible by the weakness and ignorance of the vulgar. Change is the sign of motion, and motion reveals life; if the corpse itself were dead, its decomposition would be impossible; all its composing molecules are living and working for their liberation. And you dream that the spirit is set free the first so that it may cease to live! You believe that thought and love can die when the grossest matter is imperishable! If change must be called death, we die and are reborn daily, because our forms change daily. Fear, therefore, to soil or tear your garments, but do not fear to lay them by when the hour of sleep approaches.

The embalming and mummification of bodies is a superstition which is against nature; it is an attempt to create death; it is the forcible petrification of a substance which is needed by life. But, on the other hand, we must not be quick to destroy or make away with bodies; there is no abruptness in the operations of nature, and we must not risk the violent rupture of the bonds of a departing soul. Death is never instantaneous; it is, like sleep, gradual. So long as the blood has not become absolutely cold, so long as the nerves can quiver, a man is not wholly dead, and, if none of the vital organs are destroyed, the soul can be recalled, either by accident or by a strong will. A philosopher declared that he would discredit universal testimony rather than believe in the resurrection of a dead person, but his speech was rash, for it was on the faith of universal testimony that he believed in the impossibility of resurrection. Supposing such an occurrence were proved, what would follow? Must we deny evidence or renounce reason? It would be absurd to say so. We should simply infer that we were wrong in supposing resurrection to be impossible. *Ab actu ad posse valet consecutio.*

Let us now make bold to affirm that resurrection is possible and occurs oftener than might be thought. Many persons whose deaths have been legally and scientifically attested have been subsequently found in their coffins dead indeed, but having evidently come to life and having bitten through their clenched hands so as to open the arteries and escape from their horrible agonies. A doctor would tell us that such persons were in a lethargy, and not dead. But what is lethargy? It is the name which we give to an uncompleted death, a death which is falsified by return to life. It is easy by words to escape from a difficulty when it is impossible to explain facts. The soul is joined to the body by means of sensibility, and when sensibility ceases it is a sure sign that the soul is departing. The magnetic sleep is a lethargy or factitious death which is curable at will. The etherization or torpor produced by chloroform is a real lethargy which ends sometimes in absolute death, when the soul, ravished by its temporary liberation, makes an effort of will to become free altogether, which is possible for those who have conquered hell, that is to say, whose moral strength is superior to that of astral attraction. Hence resurrection is possible only for elementary souls, and it is these above all who run the risk of involuntary revival in the tomb. Great men and true sages are never buried alive. The theory and practice of resurrection will be given in our *Ritual*; to those, meanwhile, who may ask me whether I have raised the dead, I would say that if I replied in the affirmative they would not believe me.

It now remains for us to examine whether the abolition of pain is possible, and whether it is wholesome to employ chloroform or magnetism for surgical operations. We think, and science will acknowledge it later on, that by diminishing sensibility we diminish life, and what we subtract from pain under such circumstances turns to the profit of death. Pain bears witness to the struggle for life, and hence we observe that the dressing of the wound is excessively painful in the case of persons who are operated on under anesthetics. Now, if chloroform were resorted to at each dressing, one of two things would happen – either the patient would die, or the pain would return and continue between the dressings. Nature is not violated with impunity.

21 שׁ X

DIVINATION

DENTES FURCA AMENS

THE author of this book has dared many things in his life, and never has fear retained his thought a prisoner. It is not at the same time without legitimate dread that he approaches the end of the magical doctrine. It is a question now of revealing, or rather reveiling, the Great Secret, the terrible secret, the secret of life and death, expressed in the Bible by those formidable and symbolical words of the serpent, who was himself symbolical: I. NEQUAQUAM MORIEMINI; II. SEDERITIS; III. SICUT DII; IV. SCIENTES BONUM ET MALUM. One of the privileges which belong to the initiate of the Great Arcanum, and that which sums them all, is *Divination*. According to the vulgar comprehension of the term, to divine signifies to conjecture what is unknown, but its true sense is ineffable to the point of sublimity. To divine (*divinari*) is to exercise divinity. The word *divinus* in Latin signifies something far different from *divus*, which is equivalent to the man-god. *Devin*, in French, contains the four letters of the word DIEU (God), plus the letter **N**, which corresponds in its form to the Hebrew *aleph* א, and kabbalistically and hieroglyphically expresses the Great Arcanum, of which the Tarot symbol is the figure of the Juggler. Whosoever understands perfectly the absolute numeral value of א multiplied by

N final in words which signify *science, art* or *force*, who subsequently adds the five letters of the word DEVIN, in such a way as to make five go into four, four into three, three into two, and two into one, such a person, by translating the resultant number into primitive Hebrew characters, will write the occult name of the Great Arcanum, and will possess a word of which the Sacred Tetragram itself is only the equivalent and the image.

To be a diviner, according to the force of the term, is hence to be divine, and something more mysterious still. Now, the two signs of human divinity, or of divine humanity, are prophecies and miracles. To be a prophet is to see beforehand the effects which exist in causes, to read in the astral light; to work miracles is to act upon the universal agent and subject it to our will. The author of this book will be asked whether he is a prophet and thaumaturge. Let inquirers recur to all that he wrote before certain events took place in the world; and as to anything else that he may have said or done, would anyone believe his mere word if he made any unusual statement? Furthermore, one of the essential conditions of divination is to be never constrained, never suffer temptation – in other words, being put to the test. Never have the masters of science yielded to the curiosity of anyone. The sibyls burned their books when Tarquin refused to appraise them at their proper value; the great Master was silent when He was asked for a sign of His divine mission; Agrippa perished of want rather than obey those who demanded a horoscope. To furnish proofs of science to those who suspect the very existence of the science is to initiate the unworthy, to profane the gold of the sanctuary, to deserve the excommunication of sages and the fate of betrayers.

The essence of divination, that is to say, the Great Magical Arcanum, is represented by all symbols of the science, and is connected intimately with the one and primeval doctrine of Hermes. In philosophy, it gives absolute certitude; in religion, the universal secret of faith; in physics, the composition, decomposition, recomposition, realization and adaptation of philosophical Mercury, called AZOTH by the alchemists; in dynamics it multiplies our forces by those of perpetual motion; it is at once mystical, metaphysical, and material, with correspondent effects in the three worlds; it procures charity in God, truth in science, and gold in riches, for metallic transmutation is at once an allegory and reality, as all the adepts of true science are perfectly well aware. Yes, gold can really and materially be made by means of the stone of the sages, which is an amalgam of salt, sulphur and mercury, thrice combined in AZOTH by a triple sublimation and a triple fixation. Yes, the operation is often easy, and may be accomplished in a day, an instant; at other times it requires months and years. But to succeed in the Great Work, one must be *divinus* – a diviner, in the kabbalistic sense of the term – and it is indispensable that one should have renounced, in respect of personal interest, the advantage of wealth, so as to become its dispenser. Raymund Lully enriched sovereigns, planted Europe with institutions, and remained poor. Nicholas Flamel, who, in spite of his legend, is really dead, only attained the great work when asceticism had completely detached him from riches. He was initiated by a suddenly imparted understanding of the book *Aesh Mezareph*, written in Hebrew by the kabbalist Abraham, possibly the compiler of the *Sepher Jetzirah*. Now, this understanding was, for Flamel, an intuition deserved, or rather, rendered possible, by the personal preparations of the adept. I believe that I have spoken sufficiently.

Divination is therefore an intuition, and the key of this intuition is the universal and magical doctrine of analogies. By means of these analogies, the magus interprets visions, as did the patriarch Joseph in Egypt, according to Biblical history. The analogies in the reflections of the astral light are as exact as the shades of color in the solar spectrum, and can be calculated and explained with great exactitude. It is, however, indispensable to know the dreamer's degree of intellectual life, which, indeed, he will himself reveal completely by his own dreams in a manner that will profoundly astonish himself.

Somnambulism, presentiments and second sight are simply an accidental or induced disposition to dream in a voluntary or awakened sleep – that is, to perceive the analogous reflections of the astral light, as we shall explain to demonstration in our Ritual, when providing the long-sought method of regularly producing and directing magnetic phenomena. As to divinatory instruments, they are simply a means of communication between diviner and consulter, serving merely to fix the two wills upon the same sign. Vague, complex, shifting figures help to focus the reflections of the astral fluid, and it is thus that lucidity is procured by coffee-grouts, mists, the white of egg, etc., which evoke fatidic forms, existing only in the *translucid* – that is, in the imagination of the operators. Vision in water is operated by the dazzlement and tiring of the optic nerve, which then resigns its functions to the translucid, and produces a brain illusion in which reflections of the astral light are taken for real images. Hence nervous persons, of weak sight and lively imagination, are best fitted for this species of divination, which, indeed, is most successful when performed by children. Let us not here misinterpret the function which we attribute to imagination in divinatory arts. It is by imagination assuredly that we see, and this is the natural aspect of the miracle, but we see true things, and in this consists the marvelous aspect of the natural work. We appeal to the experience of all veritable adepts. The author of this book has tested all kinds of divination, and has invariably obtained results in proportion to the exactitude of his scientific operations and the good faith of his consulters.

The Tarot, that miraculous work which inspired all the sacred books of antiquity, is, by reason of the analogical precision of its figures and numbers, the most perfect instrument of divination, and can be employed with complete confidence. Its oracles are always rigorously true, at least in a certain sense, and even when it predicts nothing it reveals secret things and gives the most wise counsel to its consulters. Alliette, who, in the last century, from a hairdresser became a kabbalist, and kabbalistically called himself Etteilla, reading his name backwards after the manner of Hebrew, Alliette, I say, after thirty years of meditation over the Tarot, was on the threshold of discovering everything that is concealed in this extraordinary work; however, he ended only by misplacing the keys, through want of their proper understanding, and inverted the order and character of the figures without, at the same time, entirely destroying their analogies, so great are the sympathy and correspondence which exist between them. The writings of Etteilla, now very rare, are obscure, wearisome and in style barbarous; they have not all been printed, and some manuscripts of this father of modern cartomancers are in the hands of a Paris bookseller who has been good enough to shew them us. Their most remarkable points are the obstinate opinions and incontestable good faith of the author, who all his life perceived the grandeur of the occult sciences, but was destined to die at the gate of the sanctuary without ever penetrating behind the veil. He had little esteem for Agrippa, made much of Jean Belot, and knew nothing of the philosophy of Paracelsus, but he possessed a highly-trained intuition, a volition most persevering, though his fancy exceeded his judgement. His endowments were insufficient for a magus and more than were needed for a skillful and accredited diviner of the vulgar order. Hence Etteilla had a fashionable success which a more accomplished magician would perhaps have been wrong to waive, but would certainly not have claimed.

When uttering at the end of our Ritual a last word upon the Tarot, we shall show the complete method of reading and hence of consulting it, not only on the probable chances of destiny, but also, and above all, upon the problems of philosophy and religion, concerning which it provides a solution which is invariably certain and also admirable in its precision, when explained in the hierarchic order of the analogy of the three worlds with the three colors and the four shades which compose the sacred septenary. All this belongs to the positive practice of magic, and can only be summarily indicated, and established theoretically in the present first part, which is concerned exclusively to the doctrine of transcendent magic, and the philosophical and religious key of the transcendent sciences, known, or rather unknown, under the name of occult.

22 ת Z

SUMMARY AND GENERAL KEY OF THE FOUR SECRET SCIENCES

SIGNA THOT PAN

LET us now summarize the entire science by its principles. Analogy is the final word of science and the first word of faith. Harmony consists in equilibrium, and equilibrium subsists by the analogy of contraries. Absolute unity is the supreme and final reason of things. Now, this reason can neither be one person nor three persons; it is a reason, and reason eminently. To create equilibrium, we must separate and unite – separate by the poles, unite by the center. To reason upon faith is to destroy faith; to create mysticism in philosophy is to assail reason. Reason and faith, by their nature, mutually exclude one another, and they unite by analogy. Analogy is the sole possible mediator between the finite and infinite. Dogma is the ever ascending hypothesis of a presumable equation. For the ignorant, it is the hypothesis which is the absolute affirmation, and the absolute affirmation which is hypothesis. Hypotheses are necessary in science, and he who seeks to realize them enlarges science without decreasing faith, for on the farther side of faith is the infinite. We believe in what we do not know, but which reason leads us to admit. To define and circumscribe the object of faith is

therefore to formulate the unknown. Professions of faith are formulations of the ignorance and aspirations of man. The theorems of science are monuments of his conquests. The man who denies God is not less fanatical than he who defines him with pretended infallibility. God is commonly defined by the enumeration of all that He is not. Man makes God by an analogy from the lesser to the greater, whence it results that the conception of God by man is ever that of an infinite man who makes man a finite god. Man can realize that which he believes in the measure of that which he knows, and by reason of that which he does not know, and he can accomplish all that he wills in the measure of that which he believes and by reason of that which he knows. The analogy of contraries is the relation of light and shade, of height and hollow, of plenum and void. Allegory, the mother of all dogmas, is the substitution of impressions for seals, of shadows for realities. It is the fable of truth and the truth of fable. One does not invent a dogma, one veils a truth, and a shade for weak eyes is produced. The initiator is not an impostor, he is a revealer, that is, following the meaning of the Latin word *revelare*, a man who veils afresh. He is the creator of a new shade.

Analogy is the key of all secrets of nature and the sole fundamental reason of all revelations. That is why religions seem to be written in the heavens and in all nature, which is just as it should be, for the work of God is the book of God, and in what He writes should be discerned the expression of His thought, and consequently of His being, since we conceive Him only as the supreme thought. Dupuis and Volney saw only a plagiarism in this splendid analogy, which should have led them to acknowledge the catholicity, that is, the universality of the primeval, one, magical, kabbalistic and immutable doctrine of revelation by analogy. Analogy yields all forces of nature to the magus; analogy is the quintessence of the philosophical stone, the secret of perpetual motion, the quadrature of the circle, the temple resting on the two pillars JAKIN and BOAZ, the key of the great arcanum, the root of the tree of life, the science of good and evil. To find the exact scale of analogies in things appreciable by science is to fix the bases of faith and thus become possessed of the rod of miracles. Now, there is a principle and a rigorous formula, which is the great arcanum. Let the wise man seek it not, since he has already found it; let the profane seek forever, they will never find it.

Metallic transmutation takes place spiritually and materially by the positive key of analogies. Occult medicine is simply the exercise of the will applied to the very source of life, to that astral light the existence of which is a fact, which has a movement conformed to calculations having the great magical arcanum for their ascending and descending scale. This universal arcanum, the final and eternal secret of transcendent initiation, is represented in the Tarot by a naked girl, who touches the earth only by one foot, has a magnetic rod in each hand, and seems to be running in a crown held up by an angel, an eagle, a bull and a lion. Fundamentally, the figure is analogous to the cherub of Jekeskiel, of which a representation is here given, and to the Indian symbol of Addhanari, which again is analogous to the ado-naï of Jekeskiel, who is vulgarly called Ezekiel. The comprehension of this figure is the key of all occult sciences. Readers of my book must already understand it philosophically if they are at all familiar with the symbolism of the Kabbalah. It remains for us now to realize what is the second and more important operation of the great work. It is something undoubtedly to find the philosophical stone, but how is it to be ground into the powder of projection? What are the uses of the magical rod? What is the real power of the divine names in the Kabbalah? The initiates know, and those who are deserving of initiation will know in turn if they discover the great arcanum by means of the very numerous and precise indications which we have given them. Why are these simple and pure truths forever and of necessity concealed? Because the elect of the understanding are always few on earth, and are encompassed by the foolish and wicked like Daniel in the den of lions. Moreover, analogy instructs us in the laws of the hierarchy, and absolute science, being an omnipotence, must be the exclusive possession of the most worthy. The confusion of the hierarchy is the actual destruction of societies, for then the blind become leaders of the blind, according to the word of the Master. Give back initiation to priests and kings and order will come forth anew. So, in my appeal to the most worthy, and in exposing myself to all the dangers and anathemas which threaten revealers, I believe myself to have done a great and useful thing, directing the breath of God living in humanity upon the social chaos, and creating priests and kings for the world to come.

A thing is not just because God wills it, but God wills it because it is just, said the angel of the schools. It is as if he said: The absolute is reason. Reason is self-existent; it is because it is, and not because we suppose it; it is or nothing is; could you wish anything to exist without reason? Madness itself does not occur without it. Reason is necessity, is law, is the rule of all liberty and the direction of all initiative. If God exists, it is by reason. The conception of an absolute God outside or independent of reason is the idol of black magic and the phantom of the fiend. The demon is death masquerading in the cast-off garments of life, the spectra of Hirrenkesept throned upon the rubbish of ruined civilizations, and concealing a loathsome nakedness by the rejected salvage of the incarnations of Vishnu.

Here ends the Doctrine of Transcendent Magic

Part II: The Ritual of Transcendent Magic

Introduction to Part II

KNOWEST thou that old queen of the world who is on the march always and wearies never? Every uncurbed passion, every selfish pleasure, every licentious energy of humanity, and all its tyrannous weakness, go before the sordid mistress of our tearful valley, and, scythe in hand, these indefatigable laborers reap their eternal harvest. That queen is old as time, but her skeleton is concealed in the wreckage of women's beauty, which she abstracts from their youth and their love. Her skull is adorned with dead tresses that are not her own. Spoliator of crowned heads, she is embellished with the plunder of queens, from the star-begemmed hair of Berenice to that, white without age, which the executioner sheared from the brow of Marie Antoinette. Her livid and frozen body is clothed in polluted garments and tattered winding-sheets. Her bony hands, covered with rings, hold diadems and chains, scepters and crossbones, jewels and ashes. When she goes by, doors open of themselves; she passes through walls; she penetrates to the cabinets of kings; she surprises the extortioners of the poor in their most secret orgies; she sits down at their board, pours out their wine, grins at their songs with her gumless teeth, takes the place of the lecherous courtesan hidden behind their curtains. She delights in the vicinity of sleeping voluptuaries; she seeks their caresses as if she hoped to grow warm in their embrace, but she freezes all those whom she touches and herself never kindles. At times, notwithstanding, one would think her seized with frenzy; she no longer stalks slowly; she runs; if her feet are too slow, she spurs a pale horse and charges all breathless through multitudes. Murder rides with her on a red charger; shaking his mane of smoke, fire flies before her with wings of scarlet and black; famine and plague follow on diseased and emaciated steeds, gleaning the few sheaves which remain to complete her harvest.

After this funeral procession come two little children, radiating with smiles and life, the intelligence and love of the coming century, the dual genius of a new-born humanity. The shadows of death fold up before them, as does night before the morning star; with nimble feet they skim the earth, and sow with full hands the hope of another year. But death will come no more, impiteous and terrible, to mow like dry grass the ripe blades of the new age; it will give place to the angel of progress, who will gently liberate souls from mortal chains, so that they may ascend to God. When men know how to live they will no longer die; they will transform like the chrysalis, which becomes a splendid butterfly. The terrors of death are daughters of ignorance, and death herself is only hideous by reason of the rubbish which covers her, and the somber hues with which her images are surrounded. Death, truly, is the birth-pang of life. There is a force in nature which dies not, and this force perpetually transforms beings to preserve them. It is the reason or word of nature. In man also there is a force analogous to that of nature, and it is the reason or word of man. The word of man is the expression of his will directed by reason, and it is omnipotent when reasonable, for it is analogous to the word of God himself. By the word of his reason man becomes conqueror of life, and can triumph over death. The entire life of man is either the parturition or miscarriage of his word. Human beings who die without having understood or formulated the word of reason, die devoid of eternal hope. To withstand successfully the phantom of death, we must be identified with the realities of life. Does it signify to God if an abortion wither, seeing that life is eternal? Does it signify to nature if unreason perish, since reason which never perishes still holds the keys of life? The first and terrible force which destroys abortions eternally was called by the Hebrews Samaël; by other easterns, Satan; and by the Latins, Lucifer. The Lucifer of the Kabbalah is not an accursed and ruined angel; he is the angel who enlightens, who regenerates by fire; he is to the angels of peace what the comet is to the mild stars of the springtime constellations. The fixed star is beautiful, radiant and calm; she drinks the celestial perfumes and gazes with love upon her sisters; clothed in her glittering robe, her forehead crowned with diamonds, she smiles as she chants her morning and evening canticle; she enjoys an eternal repose which nothing can disturb, and solemnly moves forward without departing from the rank assigned her among the sentinels of light. But the wandering comet, disheveled and of sanguinary aspect, comes hurriedly from the depths of heaven and flings herself athwart the path of the peaceful spheres, like a chariot of war between the ranks of a procession of vestals; she dares to face the burning spears of the solar guardians, and, like a bereft spouse who seeks the husband of her dreams during widowed nights, she penetrates even unto the inmost sanctuary of the god of day; again she escapes, exhaling the fires which consume her, and trailing a long conflagration behind her; the stars pale at her approach; constellate flocks, pasturing on flowers of light in the vast meadows of the sky, seem to flee before her terrible breath. The grand council of spheres assembles, and there is universal consternation; at length the loveliest of the fixed stars is commissioned to speak in the name of all the firmament and offer peace to the headlong vagabond.

"My sister," she thus commences, "why dost thou disturb the harmony of the spheres? What evil have we wrought thee? And why, instead of wandering willfully, dost thou not fix thy place like us in the court of the sun? Why dost thou not chant with us the evening hymn, clothed like ourselves in a white garment, fastened at the breast with a diamond clasp? Why float thy tresses, adrip with fiery sweat, through the mists of the night? Ah, wouldst thou but take thy place among the daughters of heaven, how much more beautiful wouldst thou be! Thy face would burn no longer with the toil of thine incredible flights; thine eyes would be pure, thy smiling countenance white and red like that of thy happy sisters; all the stars would know thee, and, far from fearing thy flight, would rejoice at thine approach; for then thou wouldst be made one with us by the indestructible bonds of universal harmony, and thy peaceful existence would be one voice more in the canticle of infinite love."

And the comet replies to the fixed star: "Believe not, O my sister, that I am permitted to wander at will and vex the harmony of the spheres! God hath appointed my path, even as thine, and if it appears to thee uncertain and rambling, it is because thy beams cannot penetrate far enough to take in the circumference of the ellipse which has been given me for my course. My fiery hair is God's beacon; I am the messenger of the suns, and I immerse myself continually in their burning heat, that I may dispense it to young worlds on my journey which have not yet sufficient warmth, and to ancient stars which have grown cold in their solitude. If I weary in my long travellings, if my beauty be less mild than thine own, and if my garments are less unspotted, yet am I a noble daughter of heaven, even as thou art. Leave me the secret of my terrible destiny, leave me the dread which surrounds me, curse me even if thou canst not comprehend; I shall none the less accomplish my work, and continue my career under the impulse of the breath of God! Happy are the stars which rest, which shine like youthful queens in the peaceful society of the universe! I am the proscribed, the eternal wanderer, who has infinity for domain. They accuse me of setting fire to the planets, the heat of which I renew; they accuse me of terrifying the stars which I enlighten; they chide me with breaking in upon universal harmony, because I do not revolve about their particular centers, because I join them one with another, directing my gaze towards the sole center of all the suns. Be reassured, therefore, O beauteous fixed star! I shall not impoverish thy peaceful light; rather I shall expend in thy service my own life and heat. I shall disappear from heaven when I shall have consumed myself, and my doom will have been glorious enough! Know that various fires burn in the temple of God, and do all give Him glory; ye are the light of golden candelabra; I am the flame of sacrifice. Let us each fulfil our destinies."

Having uttered these words, the comet tosses back her burning hair, uplifts her fiery shield, and plunges into infinite space, seeming to be lost forever.

Thus, Satan appeared and disappeared in the allegorical narratives of the Bible. "Now there was a day," says the book of Job, "when the sons of God came to present themselves before the Lord, and Satan came also among them. And the Lord said unto Satan: 'Whence comest thou?' Then Satan answered the Lord, and said: 'From going to and fro in the earth, and from walking up and down in it.'"

A Gnostic gospel, discovered in the East by a learned traveler of our acquaintance, explains the genesis of light to the profit of Lucifer, as follows: –

The self-conscious truth is the living thought. Truth is thought as it is in itself, and formulated thought is speech. When eternal thought desired a form, it said: "Let there be light." Now, this thought which speaks is the Word, and the Word said: "Let there be light," because the Word itself is the light of minds. The uncreated light, which is the divine Word, shines because it desires to be seen; when it says: "Let there be light!" it ordains that eyes shall open; it creates intelligences. When God said: "Let there be light!" Intelligence was made, and the light appeared. Now, the Intelligence which God diffused by the breath of His mouth, like a star given off from the sun, took the form of a splendid angel, who was saluted by heaven under the name of Lucifer. Intelligence awakened, and comprehended its nature completely by the understanding of that utterance of the Divine Word: "Let there be light!" It felt itself to be free because God had called it into being, and, raising up its head, with both wings extended, it replied: "I will not be slavery." "Then shalt thou be suffering," said the Uncreated Voice. "I will be liberty," replied the light. "Pride will seduce thee," said the Supreme Voice, "and thou wilt bring forth death." "I needs must strive with death to conquer life," again responded the created light. Thereupon God loosed from His bosom the shining cord which restrained the superb angel, and beholding him plunge through the night, which he furrowed with glory, He loved the offspring of His thought, and said with an ineffable smile: "How beautiful was the light!"

God has not created suffering; intelligence has accepted it to be free. And suffering has been the condition imposed upon freedom of being by Him who alone cannot err, because He is infinite. For the essence of intelligence is judgement, and the essence of judgement is liberty. The eye does not really possess light except by the faculty of closing or opening. Were it forced to be always open, it would be the slave and victim of the light, and would cease to see in order to escape the torment. Thus, created Intelligence is not happy in affirming God, except by its liberty to deny Him. Now, the Intelligence which denies, invariably affirms something, since it is asserting its liberty. It is for this reason that blasphemy glorifies God, and that hell was indispensable to the happiness of heaven. Were the light unrepelled by shadow, there would be no visible forms. If the first angels had not encountered the depths of darkness, the child-birth of God would have been incomplete, and there could have been no separation between the created and essential light. Never would Intelligence have known the goodness of God if it had never lost Him. Never would God's infinite love have shone forth in the joys of His mercy had the prodigal Son of Heaven remained in the house of His Father. When all was light, there was light nowhere; it filled the breast of God, who was laboring to bring it forth. And when He said: "Let there be light!" He permitted the darkness to repel the light, and the universe issued from chaos. The negation of the angel who, at birth, refused slavery, constituted the equilibrium of the world, and the motion of the spheres commenced. The infinite distances admired this love of liberty, which was vast enough to fill the void of eternal light, and strong enough to bear the hatred of God. But God could hate not the noblest of His children, and He proved him by His wrath only to confirm him in His power. So also the Word of God Himself, as if jealous of Lucifer, willed to come down from heaven and pass triumphantly through the shadows of hell. He willed to be proscribed and condemned; He premeditated that terrible hour when He should cry, in the throes of His agony: "My God, My God, why hast Thou forsaken Me?" As the star of the morning goes before the sun, the rebellion of Lucifer announced to new-born Nature the coming incarnation of God. Possibly Lucifer, in his fall through night, carried with him a rain of suns and stars by the attraction of his glory. Possibly our sun is a demon among the stars, as Lucifer is a star among the angels. Doubtless it is for this reason that it lights so calmly the horrible anguish of humanity and the long agony of earth -because it is free in its solitude, and possesses its light.

Such were the tendencies of the heresiarchs in the early centuries. Some, like the Ophites, adored the demon under the figure of a serpent; others, like the Cainites, justified the rebellion of the first angel like that of the first murderer. All these errors, all these shadows, all these monstrous idols of anarchy which India opposes in its symbols to the magical Trimourti, have found priests and worshippers in Christianity. The demon is nowhere mentioned in Genesis; an allegorical serpent deceives our first parents. Here is the common translation of the sacred text: "Now, the serpent was more subtle than any beast of the field which the Lord God had made." But this is what Moses says:

וְהַנָּחָשׁ הָיָה וְעָרוּם מִכֹּל חַיַּת הַשָּׂדֶה אֲשֶׁר עָשָׂה יְהוָה אֱלֹהִים:

Wha-Nahàsh haîah hâroum mi-chol hàîaht ha-shadeh asher

Hâshah Jhôah Ælohîm.

This signifies, according to the version of Fabre d'Olivet: "Now, original attraction (cupidity) was the entraining passion of all elementary life (the interior active power) of nature, the work of Jhôah, the Being of beings." But herein Fabre d'Olivet is beside the true interpretation, because he was unacquainted with the grand keys of the Kabbalah. The word Nahasch, explained by the symbolical letters of the Tarot signifies rigorously:

| 14 | נ | *Nun.* -The power which produces combinations. |
| 5 | ה | *He.* -The recipient and passive producer of forms. |

21 ש *Schin*. -The natural and central fire equilibrated by double polarization.

Thus, the word employed by Moses, read kabbalistically, gives the description and definition of that magical universal agent, represented in all theogonies by the serpent; to this agent the Hebrews applied the name of OD when it manifested its active force, of OB when it exhibited its passive force, and of AOUR when it revealed itself wholly in its equilibrated power, producer of light in heaven and gold among metals. It is therefore that old serpent which encircles the world, and places his devouring head beneath the foot of a Virgin, the type of initiation – that virgin who presents a little new-born child to the adoration of three magi, and receives from them, in exchange for this favor, gold, myrrh and frankincense. So does doctrine serve in all hieratic religions to veil the secret of those forces of nature which the initiate has at his disposal. Religious formulæ are the summaries of those words full of mystery and power which make the gods descend from heaven and become yield themselves to the will of men. Judea borrowed its secrets from Egypt; there Greece sent her hierophants, and later her theosophists, to the school of the great prophets; the Rome of the Caesars, mined by the initiation of the catacombs, collapsed one day into the Church, and a symbolism was reconstructed with the remnants of all worships which had been absorbed by the queen of the world. According to the Gospel narrative, the inscription which set forth the spiritual royalty of Christ was written in Hebrew, in Greek, and in Latin; it was the expression of the universal synthesis. Hellenism, in fact, that grand and beauteous religion of form, announced the coming of the Savior no less than the prophets of Judaism. The fable of Psyche is an ultra-Christian abstraction, and the cultus of the Pantheons, by rehabilitating Socrates, prepared altars for that unity of God, of which Israel had been the mysterious preserver. But the synagogue denied its Messiah, and the Hebrew letters were effaced, at least in the blinded eyes of the Jews. The Roman persecutors dishonored Hellenism, and it could not be restored by the false moderation of the philosopher Julian, surnamed perhaps unjustly the Apostate, since his Christianity was never sincere. The ignorance of the middle ages followed, opposing saints and virgins to gods, goddesses and nymphs; the deep sense of the Hellenic mysteries was less understood than ever; Greece herself did not only lose the traditions of her ancient cultus, but separated from the Latin Church; and thus, for Latin eyes, the Greek letters were blotted out, as the Latin letters disappeared for Greek eyes. So the inscription on the Cross of the Savior vanished entirely, and nothing except mysterious initials remained. But when science and philosophy, reconciliated with faith, shall unite all the various symbols, then shall all the magnificences of the antique worships again blossom in the memory of men, proclaiming the progress of the human mind in the intuition of the light of God. But of all forms of progress the greatest will be that which, restoring the keys of nature to the hands of science, shall enchain forever the hideous spectra of Satan, and, explaining all exceptional phenomena of nature, shall destroy the empire of superstition and idiotic credulity. To the accomplishment of this work we have consecrated our life, and do still devote it, in the most toilsome and difficult researches. We would emancipate altars by overthrowing idols; we desire the man of intelligence to become once more the priest and king of nature, and we would preserve by explanation all images of the universal sanctuary.

The prophets spoke in parables and images, because abstract language was wanting to them, and because prophetic perception, being the sentiment of harmony or of universal analogies, translates naturally by images. Taken literally by the vulgar, these images become idols or impenetrable mysteries. The sum and succession of such images and mysteries constitute what is called symbolism. Symbolism, therefore, comes from God, though it may be formulated by men. Revelation has accompanied humanity in all ages, has transfigured with human genius, but has ever expressed the same truth. True religion is one; its dogmas are simple, and within the reach of all. At the same time, the multiplicity of symbols has been a book of poesy indispensable to the education of human genius. The harmony of outward beauties and the poetry of form had to be revealed by God to the infancy of man; but soon Venus had Psyche for her rival, and Psyche enchanted Love. Thus the cultus of the form perforce yielded to those ambitious dreams which adorned already the eloquent wisdom of Plato. The advent of Christ was prepared, and for this reason was expected; it came because the world awaited it, and to become popular philosophy transformed into belief. Emancipated by this belief itself, the human mind speedily protested against the school which sought to materialize its signs, and the work of Roman Catholicism was solely the unconscious preparation for the emancipation of consciences and the establishment of the bases of universal association. All these things were the regular and normal development of divine life in humanity; for God is the great soul of all souls, the immovable center about which gravitate all intelligences like a cloud of stars.

Human understanding has had its morning; its noon will come, and the decline follow; but God will be ever the same. It seems, however, to the dwellers on the earth that the sun rises youthful and timid in the morning, shines with all its power at mid-day, and goes wearied to rest in the evening. Nevertheless, it is earth which revolves while the sun is motionless. Having faith, therefore, in human progress, and in the stability of God, the free man respects religion in its past forms, and no more blasphemes Jupiter than Jehovah; he still salutes lovingly the radiant image of the Pythian Apollo, and discovers its fraternal resemblance to the glorified countenance of the risen Redeemer. He believes in the great mission of the Catholic hierarchy, and finds satisfaction in observing the popes of the middle ages who opposed religion as a check upon the absolute power of kings; but he protests with the revolutionary centuries against the servitude of conscience which would enchain the pontifical keys; he is more protestant than Luther, since he does not even believe in the infallibility of the Augsbourg Confession, and more Catholic than the Pope, for he has no fear that religious unity will be broken by the ill-will of the courts. He trusts in God rather than Roman policy for the salvation of the unity idea; he respects the old age of the Church, but he has no fear that she will die; he knows that her apparent death will be a transfiguration and a glorious assumption.

The author of this book makes a fresh appeal to the eastern magi to come forward and recognize once again that divine Master whose cradle they saluted, the great initiator of all the ages. All His enemies have fallen; all those who condemned Him are dead; those who persecuted Him have passed into sleep for ever; He is forever alive. The envious have combined against Him, agreeing on a single point; the sectaries have united to destroy Him; they have crowned themselves kings and proscribed Him; they have become hypocrites and accused Him; they have constituted themselves judges and pronounced His sentence of death; they have turned headsmen and executed Him; they have forced Him to drink hemlock, they have crucified Him, they have stoned Him, they have burned Him and cast

His ashes to the wind; then have they turned scarlet with terror, for He still stood erect before them, impeaching them by His wounds and overwhelming them by the brightness of His scars. They believed that they had slain Him in His cradle at Bethlehem, but He is alive in Egypt! They carry Him to the summit of the mountain to cast Him down; the mob of His murderers encircles Him and already triumphs in His certain destruction; a cry is heard; is not that He who is shattered on the rocks of the abyss? They whiten and look at one another; but He, calm and smiling with pity, passes through the midst of them and disappears. Behold another mountain which they have just dyed with His blood! Behold a Cross, a sepulcher, and soldiers guarding His tomb! Madmen! The tomb is empty, and He whom they regard as dead is walking peaceably between two travelers, on the road to Emmaus. Where is He? Whither does He go? Warn the masters of the world! Tell the Cæsars that their power is threatened! By whom? By a pauper who has no stone on which to lay His head, by a man of the people condemned to the death of slaves. What insult or what madness! It matters not. The Cæsars marshal all their power; sanguinary edicts proscribe the fugitive, everywhere scaffolds rise up, circuses open arrayed with lions and gladiators, pyres are lighted, torrents of blood flow, and the Cæsars, believing themselves victorious, dare add another name to those they rehearse on their trophies; then they die, and their own apotheosis dishonors the gods whom they defended. The hatred of the world confounds Jupiter and Nero in a common contempt. Temples transformed into tombs are cast down over proscribed ashes, and above the débris of idols, above ruins of empires, He only, He whom the Cæsars proscribed, whom so many satellites pursued, whom so many executioners tortured, He only lives, alone reigns, alone triumphs!

Notwithstanding, His own disciples speedily misuse His name; pride enters the sanctuary; those who should proclaim His resurrection seek to immortalize His death, that they may feed, like ravens, on His ever-renewing flesh. In place of imitating Him by His sacrifice and shedding their blood for their children in the faith, they chain Him in the Vatican as upon another Caucasus, and become the vultures of this divine Prometheus. But what signifies their evil dream? They can only imprison His image; He Himself is free and erect, proceeding from exile to exile and from conquest to conquest; it is possible to bind a man, but not to make captive the Word of God; speech is free, and nothing can repress it. This living speech is the condemnation of the wicked, and hence they seek to destroy it, but it is they only who die, and the word of truth remains to judge their memory! Orpheus may have been rent by bacchantes, Socrates may have quaffed the poisoned cup, Jesus and His apostles have perished in the utmost tortures, John Hus, Jerome of Prague, and innumerable others, have been burned; St. Bartholomew and the massacres of September may have had in turn their victims; cossacks, knouts and Siberian deserts are still at the disposal of the Russian Emperor, but the spirit of Orpheus, of Socrates, of Jesus, and of all martyrs will live forever in the midst their dead persecutors, will stand erect amidst failing institutions and collapsing empires. It is this divine spirit, the spirit of the only Son of God, which St. John represents in his apocalypse, standing between golden candlesticks, because He is the center of all lights; having seven stars in His hand, like the seed of a new heaven; and sending down His speech upon the earth under the symbol of a two-edged sword. When the wise in their discouragement sleep through the night of doubt, the spirit of Christ is erect and vigilant. When the nations, weary of the labor which emancipates them, lie down and dream over their chains, the spirit of Christ is erect and protesting. When the blind partisans of sterilized religions cast themselves in the dust of old temples, the spirit of Christ is erect and praying. When the strong become weak, when virtues are corrupted, when all things bend and sink down in search of a shameful pasture, the spirit of Christ is erect, gazing up to heaven, and awaiting the hour of His Father.

Christ signifies priest and king by excellence. The Christ initiator of modern times came to form new priests and new kings by science and, above all, by charity. The ancient magi were priests and kings, and the Savior's advent was proclaimed to them by a star. This star was the magical pentagram, having a sacred letter at each point. It is the symbol of intelligence which rules by unity of force over the four elementary potencies; it is the pentagram of the magi, the blazing star of the children of Hiram, the prototype of equilibrated light; to each of its points a ray of light ascends, and from each a ray goes forth; it represents the grand and supreme athanor of nature, which is the body of man. The magnetic influence issues in two beams from the head, from either hand, and from either foot. The positive ray is balanced by the negative. The head corresponds with the two feet, each hand with a hand and foot, each of the two feet with the head and one hand. This ruling sign of equilibrated light represents the spirit of order and harmony; it is the sign of the omnipotence of the magus, and hence, when broken or incorrectly drawn, it represents astral intoxication, abnormal and ill-regulated projections of astral light, and, therefore, bewitchments, perversity, madness, and it is what the magi term the signature of Lucifer. There is another signature which also symbolizes the mysteries of light, namely, the sign of Solomon, whose talismans bear on one side the impression of his seal which we have given in our Doctrine, and on the other the following signature, which is the hieroglyphic theory of the composition of magnets, and represents the circulatory law of the lightning.

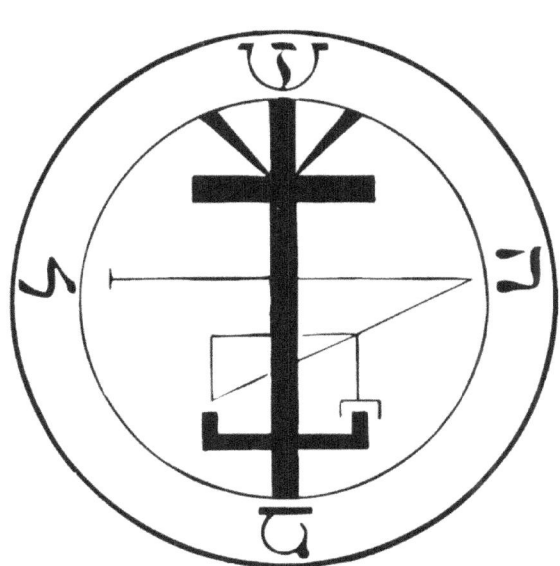

Rebellious spirits are enchained by the exhibition of the blazing five-pointed star or the seal of Solomon, because each gives them proof of their folly and threatens them with a sovereign power capable of tormenting them by their recall to order. Nothing tortures the wicked so much as goodness. Nothing is more odious to madness than reason. But if an ignorant operator should make use of these signs without knowing them, he is a blind man who discourses of light to the blind, an ass who would teach children to read.

"If the blind lead the blind," said the great and divine Hierophant, "both fall into the pot."

And now a final word to sum this entire introduction. If you be blind like Samson when you cast down the pillars of the temple, its ruins will crush you. To command nature we must be above nature by resistance of her attractions. If your mind be perfectly free from all prejudice, superstition, and incredulity, you will command spirits. If you do not obey blind forces, they will obey you. If you be wise like Solomon, you will perform the works of Solomon; if you be holy like Christ, you will accomplish the works of Christ. To direct the currents of the inconstant light, we must be established in the constant light. To command the elements, we must have overcome their hurricanes, their lightnings, their abysses, their tempests. In order to DARE we must KNOW; in order to WILL, we must DARE; we must WILL to possess empire and to reign we must BE SILENT.

Chapter I: Preparations

EVERY intention which does not assert itself by deeds is a vain intention, and the speech which expresses it is idle speech. It is action which proves life and establishes will. Hence it is said in the sacred and symbolical books that men will be judged, not according to their thoughts and their ideas, but according to their works. We must act in order to be.

We have, therefore, to treat in this place of the grand and terrific question of magical works; we are concerned no longer with theories and abstractions; we approach realities, and we are about to place the rod of miracles in the hands of the adept, saying to him at the same time: "Be not satisfied with what we tell you; act for yourself." We have to deal here with works of relative omnipotence, with the means of seizing upon the greatest secrets of nature and compelling them into the service of an enlightened and inflexible will.

Most known magical rituals are either mystifications or enigmas, and we are about to rend for the first time, after so many centuries, the veil of the occult sanctuary. To reveal the holiness of mysteries is to provide a remedy for their profanation. Such is the thought which sustains our courage and enables us to face all the perils of this enterprise, possibly the most intrepid which it has been permitted the human mind to conceive and carry out.

Magical operations are the exercise of a natural power, but one superior to the ordinary forces of nature. They are the result of a science and a practice which exalt human will beyond its normal limits. The supernatural is only the natural in an extraordinary grade, or it is the exalted natural; a miracle is a phenomenon which strikes the multitude because it is unexpected; the astonishing is that which astonishes; miracles are effects which surprise those who are ignorant of their causes, or assign them causes which are not in proportion to effects. Miracles exist only for the ignorant, but, as there is scarcely any absolute science among men, the supernatural can still obtain, and does so indeed for the whole world. Let us set out by saying that we believe in all miracles because we are convinced and certain, even from our own experience, of their entire possibility. There are some which we do not explain, though we regard them as no less explicable. From the greater to the lesser, from the lesser to the greater, the consequences are identically related and the proportions progressively rigorous. But in order to work miracles we must be outside the normal conditions of humanity; we must either be abstracted by wisdom or exalted by madness, either superior to all passions or beyond them through ecstasy or frenzy. Such is the first and most indispensable preparation of the operator. Hence, by a providential or fatal law, the magician can only exercise omnipotence in inverse proportion to his material interest; the alchemist makes so much the more gold as he is the more resigned to privations, and the more esteems that poverty which protects the secrets of the *magnum opus*. Only the adept whose heart is passionless will dispose of the love and hate of those whom he would make instruments of his science; the myth of Genesis is eternally true, and God permits the tree of knowledge to be approached only by those men who are sufficiently strong and self-denying not to covet its fruits. Ye, therefore, who seek in science a means to satisfy your passions, pause in this fatal way; you will find nothing but madness or death. This is the meaning of the vulgar tradition that the devil ends sooner or later by strangling sorcerers. The Magus must be impassible, sober and chaste, disinterested, impenetrable, and inaccessible to any kind of prejudice or terror. He must be without bodily defects, and proof against all contradictions and all difficulties. The first and most important of magical operations is the attainment of this rare pre-eminence.

We have said that impassioned ecstasy may produce the same results as absolute superiority, and this is true as to the issue, but not as to the direction of magical operations. Passion forcibly projects the astral light and impresses unforeseen movements on the universal agent, but it cannot check with the facility that it impels, and its destiny then resembles that of Hippolytus dragged by his own horses, or Phalaris victimized himself by the instrument of torture which he had invented for others. Human volition realized by action is like a cannon-ball, and recedes before no obstacle. It either passes through it or is buried in it, but if it advance with patience and perseverance, it is never lost; it is like the wave which returns incessantly and wears away iron in the end.

Man can be modified by habit, which becomes, according to the proverb, his second nature. By means of persevering and graduated athletics, the powers and activity of the body can be developed to an astonishing extent. It is the same with the powers of the soul. Would you reign over yourselves and others? Learn how to will. How can one learn to will? This is the first arcanum of magical initiation, and it was to make it understood fundamentally that the ancient depositaries of priestly art surrounded the approaches of the sanctuary with so many terrors and illusions. They did not believe in a will until it had produced its proofs, and they were right. Power is justified by victories. Indolence and forgetfulness are enemies of will, and for this reason all religions have multiplied their observances and made their worship minute and difficult. The more we restrain ourselves for an idea, the greater is the strength we acquire within the scope of that idea. Are not mothers more partial to the children who have caused them most suffering and cost them most anxieties? So does the power of religions reside exclusively in the inflexible will of those who practice them. So long as there is one faithful person to believe in the Holy Sacrifice of the Mass, there will be a priest to celebrate it for him; and so long as there is a priest who daily recites his breviary, there will be a pope in the world. Observances, apparently most insignificant and most foreign in themselves to the proposed end, lead, notwithstanding, to that end by education and exercise of will. If a peasant rose up every morning at two or three o'clock, and went daily a long distance from home to gather a sprig of the same herb before the rising of the sun, he would be able to perform a great number of prodigies by merely carrying this herb upon his person, for it would be the sign of his will, and would become by his will itself all that he required it to become in the interest of his desires. In order to do a thing, we must believe in our possibility of doing it, and this faith must forthwith be translated into acts. When a child says: "I cannot," his mother answers: "Try." Faith does not even try; it begins with the certitude of completing, and it proceeds calmly, as if omnipotence were at its disposal and eternity before it. What seek you, therefore, from the science of the magi? Dare to formulate your desire, then set to work at once, and do not cease acting after the same manner and for the same end; what you will shall come to pass, and for you and by you it has indeed already begun. Sixtus V said, while watching his flocks: "I desire to be pope." You are a beggar, and you desire to make gold; set to work and never leave off. I promise you, in the name of science, all the treasures of Flamel and Raymund Lully. "What is the first thing to do?" Believe in your power, then act. "But how act?" Rise daily at the same hour, and that early; bathe at a spring before daybreak, and in all seasons; never wear soiled clothes, rather wash them yourself if needful; practice voluntary privations, that you may be better able to bear those which come without seeking; then silence every desire which is foreign to the fulfilment of the great work.

"What! By bathing daily in a spring, I shall make gold?" You will work in order to make it. "It is a mockery!" No, it is an arcanum. "How can I make use of an arcanum which I fail to understand?" Believe and act; you will understand later.

One day a person said to me: "I would that I could be a fervent Catholic, but I am a Voltairean. What would I not give to have faith!" I replied: "Say 'I would' no longer; say 'I will', and I promise you that you will believe. You tell me that you are a Voltairean, and of all the various presentations of faith that of the Jesuits is most repugnant to you, but at the same time seems the most powerful and desirable. Perform the exercises of St. Ignatius again and again, without allowing yourself to be discouraged, and you will attain the faith of a Jesuit. The result is infallible, and should you then have the simplicity to ascribe it to a miracle, you deceive yourself now in thinking that you are a Voltairean."

An idle man will never become a magician. Magic is an exercise of all hours and all moments. The operator of great works must be absolute master of himself; he must know how to conquer the allurements of pleasure, appetite and sleep; he must be insensible to success and to indignity. His life must be that of a will directed by one thought, and served by entire nature, which he will have made subject to mind in his own organs, and by sympathy in all the universal forces which are their correspondents. All faculties and all senses should share in the work; nothing in the priest of Hermes has the right to remain idle; intelligence must be formulated by signs and summarized by characters or pantacles; will must be determined by words, and must fulfil words by deeds; the magical idea must be rendered into light for the eyes, harmony for the ears, perfumes for the sense of smell, savors for the palate, objects for the touch; the operator, in a word, must realize in his whole life that which he wishes to realize in the world without him; he must become a *magnet* to attract the desired thing; and when he shall be sufficiently magnetic, let him be assured that the thing will come of itself, and without thinking of it.

It is important for the magus to be acquainted with the secrets of science, but he may know them by intuition, and without formal learning. Solitaries, living in the habitual contemplation of nature, frequently divine her harmonies, and are more instructed in their simple good sense than doctors, whose natural discernment is falsified by the sophistries of the schools. True practical magicians are almost invariably found in the country, and are frequently uninstructed persons and simple shepherds. Furthermore, certain physical organizations are better adapted than others for the revelations of the occult world; there are sensitive and sympathetic natures, with whom intuition in the astral light is, so to speak, inborn; certain afflictions and certain complaints modify the nervous system and, independently of the concurrence of the will, may convert it into a divinatory apparatus of less or more perfection; but these phenomena are exceptional, and generally magical power should, and can, be acquired by perseverance and labor. There are also some substances which produce ecstasy, and dispose towards the magnetic sleep; there are some which place at the service of imagination all the most lively and highly colored reflections of the elementary light; but the use of such things is dangerous, for they commonly occasion stupefaction and intoxication. They are used, notwithstanding, but in carefully calculated quantities, and under wholly exceptional circumstances.

He who decides to devote himself seriously to magical works, after fortifying his mind against all danger of hallucination and fright, must purify himself without and within for forty days. The number forty is sacred, and its very figure is magical. In Arabic numerals it consists of the circle, which is the type of the infinite, and of the 4, which sums the triad by unity. In Roman numerals, arranged after the following manner, it represents the sign of the fundamental doctrine of Hermes, and the character of the Seal of Solomon:

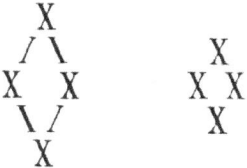

The purification of the Magus consists in the renunciation of coarse enjoyments, in a temperate and vegetable diet, in refraining from intoxicating drink, and in regulating the hours of sleep. This preparation has been indicated and represented in all forms of worship by a period of penitence and trials preceding the symbolical feasts of life-renewal.

As already stated, the most scrupulous external cleanliness must be observed; the poorest person can find spring water. All clothes, furniture and vessels made use of must also be carefully washed, whether by ourselves or others. All dirt is evidence of negligence, and negligence is deadly in magic. The atmosphere must be purified at rising and retiring with a perfume composed of the juice of laurels, salt, camphor, white resin, and sulphur, repeating at the same time the four sacred names, while turning successively towards the four cardinal points. We must divulge to no one the works that we accomplish, for, as already said in our Doctrine, mystery is the exact and essential condition of all the operations of science. The inquisitive must be misled by the pretense of other occupations and other researches, such as chemical experiments for industrial purposes, hygienic prescriptions, the investigation of some natural secrets, and so on; but the forbidden name of magic must never be pronounced.

The magus must be isolated at the beginning and difficult to approach, so that he may concentrate his power and select his points of contact, but in proportion as he is austere and inaccessible at first, so will he be popular and sought after when he shall have magnetized his chain and chosen his place in a current of ideas and of light. A laborious and poor existence is so favorable to practical initiation that the greatest masters have preferred it, even when the wealth of the world was at their disposal. Then it is that Satan, that is, the spirit of ignorance, who scorns, suspects, and detests science because at heart he fears it, comes to tempt the future master of the world by saying to him: "If thou art the Son of God, command these stones to become bread." Then it is that mercenary men seek to humiliate the prince of knowledge by perplexing, depreciating, or sordidly exploiting his labor, the slice of bread that he deigns to need is broken into ten fragments, so that he may ten times stretch forth his hand. But the magus does not even smile at the absurdity, and calmly pursues his work.

So far as may be possible, we must avoid the sight of hideous objects and uncomely persons, must decline eating with those whom we do not esteem, and must live in the most uniform and studied manner. We must hold ourselves in the highest respect, and consider that we are dethroned sovereigns who consent to existence in order to reconquer our crowns. We must be mild and considerate to all, but in social relations must never permit ourselves to be absorbed, and must withdraw from circles in which we cannot acquire some initiative. Finally, we may and should fulfil the duties and practice the rites of the cultus to which we belong. Now, of all forms of worship the most magical is that which most realizes the miraculous, which bases the most inconceivable mysteries upon the highest reasons, which has lights equivalent to its shadows, which popularizes miracles, and incarnates God in all mankind by faith. This religion has existed always in the world, and under many names has been ever the one and ruling religion. It has now among the nations of the earth three apparently hostile forms, which are, however, destined to unite before long for the constitution of a universal Church. I refer to the Greek orthodoxy, Roman Catholicism, and a final transfiguration of the religion of Buddha.

We have now made it plain, as we believe, that our magic is opposed to the goëtic and necromantic kinds; it is at once an absolute science and religion, which should not indeed destroy and absorb all opinions and all forms of worship, but should regenerate and direct them by reconstituting the circle of initiates, and thus providing the blind masses with wise and clear-seeing leaders.

We are living at a period when nothing remains to destroy and everything to remake. "Remake what? The past?" No one can remake the past. "What, then, shall we reconstruct? Temples and thrones?" To what purpose, since the former ones have been cast down? "You might as well say: my house has collapsed from age, of what use is it to build another?" But will the house that you contemplate erecting be like that which has fallen? No, for the one was old and the other will be new. "Notwithstanding, it will be always a house." What else can you wish?

Chapter II: Magical Equilibrium

EQUILIBRIUM is the consequence of two forces. If two forces are absolutely and invariably equal, the equilibrium will be immobility, and therefore the negation of life. Movement is the result of an alternate preponderance. The impulsion given to one of the sides of a balance necessarily determines the motion of the other. Thus, contraries act on one another, throughout all Nature, by correspondence and analogical connection. All life is composed of an aspiration and a respiration; creation is the assumption of a shadow to serve as a bound for light, of a void to serve as space for the plenitude, of a passive fructified principle to sustain and realize the power of the active generating principle. All nature is bisexual, and the movement which produces the appearances of death and life is a continual generation. God loves the void which He made in order to fill it; science loves the ignorance which it enlightens; strength loves the weakness which it supports; good loves the apparent evil which glorifies it; day is desirous of night, and pursues it unceasingly round the world; love is at once a thirst and a plenitude which must diffuse itself. He who gives receives, and he who receives gives; movement is a continual interchange. To know the law of this change, to be acquainted with the alternate or simultaneous proportion of these forces, is to possess the first principles of the great magical arcanum, which constitutes true human divinity. Scientifically, we can appreciate the various manifestations of the universal movement through electric or magnetic phenomena. Electrical apparatuses above all reveal materially and positively the affinities and antipathies of certain substances. The marriage of copper with zinc, the action of all metals in the galvanic pile, are perpetual and unmistakable revelations. Let physicists seek and find out; ever will the kabbalists explain the discoveries of science!

The human body is subject, like the earth, to a dual law; it attracts and it radiates; it is magnetized by an androgyne magnetism, and reacts on the two powers of the soul, the intellectual and sensitive, but in proportion to the alternating preponderances of the two sexes in their physical organism. The art of the magnetizer consists wholly in the knowledge and use of this law. To polarize the action and impart to the agent a bisexual and alternated force is the method still unknown and sought vainly for directing the phenomena of magnetism at will, but tact most experienced and great precision in the interior movements are required to prevent the confusion of the signs of magnetic aspiration with those of respiration; we must also be perfectly acquainted with occult anatomy and the special temperament of the persons on whom we are operating. Bad faith and bad will in subjects constitute the gravest hindrance to the direction of magnetism. Women above all – who are essentially and invariably actresses, who take pleasure in impressing others so that they may impress themselves, and are themselves the first to be deceived when playing their neurotic melodramas – are the true black magic of magnetism. So is it forever impossible that magnetizers who are uninitiated in the supreme secrets, and unassisted by the lights of the Kabbalah, should govern this refractory and fugitive element. To be master of woman, we must distract and deceive her skillfully by allowing her to suppose that it is she who is deceiving us. This advice, which we offer chiefly to magnetizing physicians, might find its place and application in conjugal polity.

Man can produce two breathings at his pleasure, one warm and the other cold; he can project also either active or passive light at will; but he must acquire the consciousness of this power by dwelling habitually thereon. The same manual gesture may alternately aspire and respire what we are accustomed to call the fluid, and the magnetizer will himself be warned of the result of his intention by an alternative sensation of warmth and cold in the hand, or in both hands when both are being used, which sensation the subject should experience at the same time, but in a contrary sense, that is, with a wholly opposite alternative.

The pentagram, or sign of the microcosmos, represents, among other magical mysteries, the double sympathy of the human extremities with each other and with the circulation of the astral light in the human body. Thus, when a man is represented in the star of the pentagram, as may be seen in the "Occult Philosophy" of Agrippa, it should be observed that the head corresponds in masculine sympathy with the right foot and in feminine sympathy with the left foot; that the right hand corresponds in the same way with the left hand and left foot, and reciprocally of the other hand. This must be borne in mind when making magnetic passes, if we seek to govern the whole organism and bind all members by their proper chains of analogy and natural sympathy. The same knowledge is necessary for the use of the pentagram in the conjuration of spirits, and in the evocation of errant spirits in the astral light, vulgarly called necromancy, as we shall explain in the fifth chapter of this Ritual. But it is well to observe here that every action promotes a reaction, and that in magnetizing others, or influencing them magically, we establish between them and ourselves a current of contrary but analogous influence which may subject us to them instead of subjecting them to us, as happens frequently enough in those operations which have the sympathy of love for their object. Hence it is highly essential to be on our defense while we are attacking, so as not to aspire on the left while we respire on the right. The magical androgyne depicted in the frontispiece of the Ritual has SOLVE inscribed upon the right and COAGULA on the left arm, which corresponds to the symbolical figure of the architects of the second temple, who bore the sword in one hand and their trowel in the other. While building they had also to defend their work and disperse their enemies; nature herself does likewise, destroying and regenerating at the same time. Now, according to the allegory of Duchentau's Magical Calendar, man, that is to say, the initiate, is the ape of nature, who confines him by a chain, but makes him act unceasingly, imitating the proceedings and works of his divine mistress and imperishable model.

The alternate use of contrary forces, warmth after cold, mildness after severity, love after anger, etc., is the secret of perpetual motion and the permanence of power; coquettes feel this instinctively, and hence they make their admirers pass from hope to fear, from joy to despondency. To operate always on the same side and in the same manner is to overweight one plate of the balance, and complete destruction of equilibrium is the speedy result. Continual caressings beget satiety, disgust and antipathy, just as constant coldness and severity in the long run alienate and discourage affection. An unvarying and ardent fire in alchemy calcines the first matter and not seldom explodes the hermetic vessel; the heat of lime and mineral manure must be substituted at regular intervals for the heat of flame. And so also in magic; the works of wrath or severity must be tempered by those of beneficence and love, and if the will of the operator be always at the same tension and directed along the same line, great weariness will ensue, together with a species of moral impotence.

Thus, the magus should not live altogether in his laboratory, among his athanors, elixirs and pantacles. However devouring be the glance of that Circe who is called occult power, we must be able to confront her on occasion with the sword of Ulysses, and resolutely withdraw our lips for a time from the chalice which she offers us. A magical operation should always be followed by a rest of equal length and a distraction analogous but contrary in its object. To strive continually against nature in order to her rule and conquest is to risk reason and life. Paracelsus dared to do so, but even in the warfare itself he employed equilibrated forces and opposed the intoxication of wine to that of intelligence. So was Paracelsus a man of inspiration and miracles; yet his life was exhausted by this devouring activity, or rather its vestment was rapidly rent and worn out; but men like Paracelsus use and abuse fearlessly; they know that they can no more die than grow old here below.

Nothing induces us towards joy so effectually as sorrow; nothing is nearer to sorrow than joy. Hence the uninstructed operator is astounded by attaining the very opposite of his proposed results, because he does not know how to cross or alternate his action; he seeks to bewitch his enemy, and himself becomes ill and miserable; he desires to make himself loved, and he consumes himself for women who deride him; he endeavors to make gold, and he exhausts all his resources; his torture is that of Tantalus eternally; ever does the water flow back when he stoops down to drink. The ancients in their symbols and magical operations multiplied the signs of the duad, so that its law of equilibrium might be remembered. In their evocations they constructed two altars, and immolated two victims, one white and one black; the operator, whether male or female, holding a sword in one hand and a wand in the other, had one foot shod and the other bared. At the same time, either one or three persons were required for magical works, because the duad would mean immobility or death in the absence of an equilibrating motor; and when a man and a woman participated in the ceremony, the operator was either a virgin, a hermaphrodite, or a child. I shall be asked whether the eccentricity of these rites is arbitrary, and whether it's one end is the exercise of the will by the mere multiplication of difficulties in magical work? I answer that in magic there is nothing arbitrary, because everything is ruled and predetermined by the one and universal dogma of Hermes, that of analogy in the three worlds. Each sign corresponds to an idea, and to the special form of an idea; each act expresses a volition corresponding to a thought, and formulates the analogies of that thought and that will. The rites are, therefore, prearranged by the science itself. The uninstructed person who is not acquainted with the three powers is subject to their mysterious fascination; the sage understands those powers, and makes them the instrument of his will, but when they are accomplished with exactitude and faith, they are never ineffectual.

All magical instruments must be duplicated; there must be two swords, two wands, two cups, two chafing-dishes, two pantacles and two lamps; two vestments must be worn, one over the other, and they must be of contrary colors, a rule still followed by Catholic priests; and either no metal, or two at the least, must be worn. The crowns of laurel, rue, mugwort or vervain must, in like manner, be double; one of them is used in evocations, while the other is burnt, the crackling which it makes and the curls of the smoke which it produces being observed like an augury. Nor is the observance vain, for in the magical work all instruments of art are magnetized by the operator; the air is charged with his perfumes, the fire which he has consecrated is subject to his will, the forces of nature seem to hear and answer him; he reads in all forms the modifications and complements of his thought. He perceives the water agitated, and, as it were, bubbling of itself, the fire blazing up or extinguishing suddenly, the leaves of garlands rustling, the magical rod moving spontaneously, and strange, unknown voices passing through the air. It was in such evocations that Julian beheld the beloved phantoms of his dethroned gods, and was appalled at their decrepitude and pallor.

I am aware that Christianity has forever suppressed ceremonial magic, and that it severely proscribes the evocations and sacrifices of the old world. It is not, therefore, our intention to give a new ground for their existence by revealing their antique mysteries after the lapse of so many centuries. Even in this very order of phenomena, our experiments have been scholarly researches and nothing more. We have confirmed facts that we might appreciate causes, and it has not been our pretension to restore rites which are forever destroyed. The orthodoxy of Israel, that religion which is so rational, so divine and so ill known, condemns, no less than Christianity, the mysteries of ceremonial magic. From the standpoint of the tribe of Levi, the exercise of transcendent magic must be considered as a usurpation of the priesthood; and the same reason has caused the proscription of operative magic by every official cultus. To demonstrate the natural foundation of the marvelous, and to produce it at will, is to annihilate for the vulgar mind that convincing evidence from miracles which is claimed by each religion as its exclusive property and its final argument. Respect for established religions, but room also for science! We have passed, thank God, the days of inquisitions and pyres; unhappy men of learning are no longer murdered on the faith of a few distraught fanatics or hysterical girls. For the rest, let it be clearly understood that our undertaking is concerned with studies of the curious and not with an impossible propaganda. Those who may blame us for daring to term ourselves magician have nothing to fear from the example, it being wholly improbable that they will ever become sorcerers.

Chapter III: The Triangle of Pantacles

THE Abbot Trithemius, who in magic was the master of Cornelius Agrippa, explains, in his "Steganography," the secret of conjurations and evocations after a very natural and philosophical manner, though possibly, for that very reason, too simply and too easily. He tells us that to evoke a spirit is to enter into the dominant thought of that spirit, and if we raise ourselves morally higher along the same line, we shall draw the spirit away with us, and it will certainly serve us. To conjure is to oppose the resistance of a current and a chain to an isolated spirit – *cum jurare*, to swear together, that is, to make a common act of faith. The greater the strength and enthusiasm of this faith, the more efficacious is the conjuration. This is why new-born Christianity silenced the oracles; it alone possessed inspiration, its only force. Later on, when St. Peter grew old, that is, when the world believed that it had a legal case against the Papacy, the spirit of prophecy came to replace the oracles. Savonarola, Joachim of Flores, John Hus and so many others influenced by turns the minds of men, and interpreted, by lamentations and menaces, the secret anxieties and rebellions of all hearts.

We may act individually when evoking a spirit, but to conjure we must speak in the name of a circle or an association; this is the significance of the hieroglyphical circle inscribed round the magus who is operating, and out of which he must not pass unless he wishes at the same moment to be stripped of all his power. Let us grapple at this point with the vital and palmary question, whether the real evocation and real conjuration of spirits are things possible, and whether such possibility can be scientifically demonstrated. To the first part of the question it may be replied out of hand that everything which is not an evident impossibility can and must be admitted as provisionally possible. As to the second part, we affirm that in virtue of the great magical dogma of the hierarchy and of universal analogy, the kabbalistic possibility of real evocations can be demonstrated; concerning the phenomenal reality consequent upon magical operations accomplished with sincerity, this is a matter of experience; as already described, we have established it in our own persons, and by means of this Ritual we shall place our readers in a position to renew and confirm our experiences.

Nothing in nature perishes; whatsoever has lived goes on living always under new forms; but even the anterior forms are not destroyed, since they remain in our memory. Do we not still see in imagination the child whom we once knew, though now he is an old man? The very traces which we believe to be effaced from our memory are not in reality blotted out, for a fortuitous circumstance may evoke and recall them. But after what manner do we see them? As we have already said, it is in the astral light, which transmits them to our brain by the mechanism of the nervous system. On the other hand, all forms are proportional and analogical to the idea which has determined them; they are the natural character, the signature of that idea, as the magi term it, and so soon as the idea is actively evoked, the form is realized and bodied forth. Schrœpffer, the famous illuminé of Leipzig, terrified all Germany with his evocations, and his audacity in magical experiments was so great that his reputation became an insupportable burden. He allowed himself to be carried away by the immense current of hallucinations which he had produced; the visions of the other world disgusted him with this, and he killed himself. His story should be a warning to those who are fascinated by ceremonial magic. Nature is not outraged with impunity, and no one can play safely with unknown and incalculable forces. It is this consideration which has led, and will ever lead us, to deny the vain curiosity of those who would see in order that they may believe, and we reply to them in the same words as we replied to an eminent Englishman who threatened us with his skepticism: "You are perfectly within your right in refusing to believe: for our own part, it will not make us more discouraged or less convinced." To those who may assure us that they have scrupulously and boldly fulfilled all the rites, and that there has been no result, we would recommend that they should stay their hand, as it is possibly a warning of nature, who will not lend herself to them for these anomalous works; but if they persist in their curiosity, they have only to start afresh.

The triad, being the foundation of magical doctrine, must necessarily be observed in evocations; so it is the symbolical number of realization and effect. The letter ש is commonly traced upon kabbalistic pantacles which have the fulfilment of a desire for their object. It is also the sign of the scapegoat in mystic kabbalah, and Saint Martin observes that inserted in the incommunicable tetragram it forms the name of the Redeemer. יהשוה. It is this which the mystagogues of the middle ages represented in their nocturnal assemblies by the exhibition of a symbolical goat, carrying a lighted torch between its two horns. In the fifteenth chapter of this Ritual we shall describe the allegorical forms and strange cultus of this monstrous animal, which represented nature doomed to anathema but ransomed by the sign of light. The gnostic agapæ and pagan priapic orgies which followed in its honor sufficiently revealed the moral consequence which the adepts drew from the exhibition. All this will be explained, together with the rites, decried and now regarded as fabulous, of the great Sabbath and of Black Magic.

Within the grand circle of evocations a triangle was usually traced, and the side towards which the upper point should be directed was a matter for careful observation. If the spirit were supposed to be from heaven, the operator placed himself at the top, and set the altar of fumigations at the bottom; but if the spirit came from the abyss this method was reversed. Moreover, the sacred symbol of two interlaced triangles, forming the six-pointed star, known in magic as the pantacle or Seal of Solomon, must be worn upon the forehead and the breast, and graven in the right hand. Independently of these signs, the ancients, in their evocations, made use of those mystical combinations of divine names which we have reproduced in our Doctrine from the Hebrew kabbalists. The magic triangle of pagan theosophists was the celebrated ABRACADABRA, to which they attributed extraordinary virtues, and represented as follows:

ABRACADABRA
ABRACADABR
ABRACADAB
ABRACADA
ABRACAD
ABRACA
ABRAC
ABRA
ABR
AB
A

This combination of letters is a key of the pentagram. The initial A is repeated five and reproduced thirty times, thus giving the elements and numbers of the two following figures:

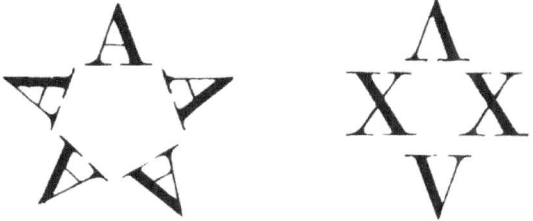

The isolated A represents the unity of the first principle, otherwise, the intellectual or active agent. A united to B represents the fertilization of the duad by the monad. R is the sign of the triad, because it represents hieroglyphically the emission which results from the union of the two principles. The number 11, which is that of the letters of the word, combines the unity of the initiate with the denary of Pythagoras, and the number 66, the added total of all the letters, form kabbalistically the number 12, which is the square of the triad, and consequently the mystic quadrature of the circle. We may remark, in passing, that the author of the Apocalypse, that key of the Christian Kabbalah, composed the number of the beast, that is to say, of idolatry, by adding a 6 to the double senary of ABRACADABRA, which gives 18 kabbalistically, the number attributed in the Tarot to the hieroglyphic sign of night and of the profane – the moon, together with the towers, dog, wolf and crab – a mysterious and obscure number, the kabbalistic key of which is 9, the number of initiation. On this subject the sacred kabbalist says expressly: "He that hath understanding (that is, the key of kabbalistic numbers), let him count the number of the beast, for it is the number of a man, and the number of him is 666." It is, in fact, the decade of Pythagoras multiplied by itself and added to the sum of the triangular Pantacle of Abracadabra; it is thus the sum of all magic in the ancient world, the entire programme of human genius, which is the divine genius of the Gospel sought to absorb or transplant.

These hieroglyphical combinations of letters and numbers belong to the practical part of the kabbalah, which, from this point of view, is divided into Gematriah and Temurah. Such calculations, which now seem to us arbitrary or devoid of interest, then belonged then to the philosophical symbolism of the East, and were of the highest importance in the teaching of holy things emanating from the occult sciences. The absolute kabbalistic alphabet, which connected primitive ideas with allegories, allegories with letters, and letters with numbers, was then called the keys of Solomon. We have stated already that these keys, preserved to our own day, but wholly misconstrued, are nothing else than the game of Tarot, the antique allegories of which were remarked and appreciated for the first time in the modern world by the learned archaeologist, Court de Gebelin.

The double triangle of Solomon is explained by St. John in a remarkable manner. He says, "There are three which give testimony in heaven – the Father, the Word and the Holy Spirit; and there are three which give testimony on earth – the spirit, the water and the blood."

Thus, St. John agrees with the masters of Hermetic philosophy, who attribute to their sulphur the name of ether, to their mercury that of philosophical water, and to their salt the qualification of the dragon's blood or menstruum of the earth; blood or salt corresponds by opposition with the Father, azotic or mercurial water with the Word or Logos, and the ether with the Holy Spirit. But the things of transcendent symbolism can only be understood rightly by the true children of science.

The threefold repetition of names with varied intonations were united to triangular combinations in magical ceremonies. The magic rod was frequently surmounted with a small magnetized fork, which Paracelsus replaced by the trident represented below.

This trident is a pantacle expressing the synthesis of the triad in the monad, thus completing the sacred tetrad. He ascribed to this figure all the virtues which kabbalistic Hebrews attribute to the name of Jehovah, and the thaumaturgic properties of the Abracadabra, used by the hierophants of Alexandria. Let us recognize here that it is a pantacle, and consequently a concrete and an absolute sign of an entire doctrine, which has been that of an immense magnetic circle, not only for ancient philosophers, but also for adepts of the middle ages. The restoration in our own day of its original value by the comprehension of its mysteries, might not that also restore all its miraculous virtue and all its power against human diseases?

The old sorceresses, when they spent the night at the meeting-place of three cross-roads, yelled three times in honor of triple Hecate. All these figures, all these dispositions of numbers and of characters, are, as we have said, so many instruments for the education of the will, by fixing and determining its habits. They serve, furthermore, to conjoin all the powers of the human soul in action, and to increase the creative force of imagination; it is the gymnastics of thought in training for realization; so, the effect of these practices is infallible, like nature, when they are fulfilled with absolute confidence and indomitable perseverance. The Grand Master tells us that faith could transplant trees into the sea and remove mountains. Even a superstitious and insensate practice is efficacious because it is a realization of will. Hence a prayer is more powerful if we go to church to say it than when it is said at home, and it will work miracles if we fare to a famous sanctuary for the purpose, in other words, to one which is strongly magnetized by the enormous number of its frequenters, traversing two or three hundred leagues with bare feet and asking alms by the way. Men laugh at the simple woman who denies herself a pennyworth of milk in the morning that she may carry a penny taper to burn on the magic triangle in a chapel; but they who laugh are ignorant, and the simple woman does not pay too dearly for what she thus purchases of resignation and of courage. Great minds with great pride pass by, shrugging their shoulders; they rise up against superstition with a din which shakes the world; and what happens? The towers of great minds topple over, and their ruins revert to the providers and purchasers of penny tapers, who are content to hear it proclaimed everywhere that their reign is forever ended, provided that they rule always.

The great religions have never had more than one serious rival, and this rival is magic. Magic produced the occult associations which brought about the revolution termed the Renaissance; but it has been the doom of the human mind, blinded by insensate passions, to realize literally the allegorical history of the Hebrew Hercules: by overthrowing the pillars of the temple, it has itself been buried under the ruins. The masonic associations of the present time are no less ignorant of the high meaning of their symbols than are the rabbins of the Sepher Jetzirah and the Zohar upon the ascending scale of the three degrees, with the transverse progression from right to left and from left to right of the kabbalistic septenary. The compass of the G∴ A∴ and the square of Solomon have become the gross and material level of unintelligent Jacobinism, realized by a steel triangle; this obtains both for heaven and earth. The initiated divulgers to whom the illuminated Cazotte predicted a violent death have, in our own days, exceeded the sin of Adam; having rashly gathered the fruits of the tree of knowledge, which they did not know how to use for their nourishment, they have cast it to the beasts and reptiles of the earth. So was the reign of superstition inaugurated, and it must persist until the period when true religion shall be again constituted on the eternal foundations of the hierarchy of three degrees, and of the triple power which the hierarchy exercises blindly or providentially in the three worlds.

Chapter IV: The Conjuration of the Four

THE four elementary forms separate and distinguish the created spirits which the universal movement disengages from the central fire. The spirit everywhere toils and fructifies matter by life; all matter is animated; thought and soul are everywhere. By possessing ourselves of the thought which produces diverse forms, we become the master of forms and make them serve our purposes. The astral light is saturated with such souls, which it disengages in the unceasing generation of beings. These souls have imperfect wills, which can be governed and employed by more powerful wills; then great invisible chains form and may occasion or determine great elementary commotions. The phenomena established by the criminal trials of magic, and quite recently by M. Eudes de Mirville, have no other cause. Elementary spirits are like children; they chiefly torment those who trouble about them, unless, indeed, they are controlled by high reason and great severity. We designate these spirits under the name of occult elements, and it is these who frequently occasion our bizarre or disturbing dreams, who produce the movements of the divining rod and rappings upon walls or furniture, but they can manifest no thought other than our own, and when we are not thinking, they speak to us with all the incoherence of dreams. They reproduce good and evil indifferently, for they are without free will, and are hence irresponsible; they exhibit themselves to ecstatics and somnambulists under incomplete and fugitive forms. This explains the nightmares of St. Anthony, and most probably the visions of Swedenborg. Such creatures are neither damned nor guilty, they are curious and innocent. We may use or abuse them like animals or children. Therefore, the magus who makes use of them assumes a terrible responsibility, for he must expiate all the evil which he causes them to accomplish, and the intensity of his punishment will be in proportion to that of the power which he may have exercised by their mediation.

To govern elementary spirits, and thus become the king of the occult elements, we must first have undergone the four ordeals of ancient initiations; and seeing that these initiations exist no longer, we must have substituted analogous experiences, such as exposing ourselves boldly in a fire, crossing an abyss by means of the trunk of a tree or a plank, scaling a perpendicular mountain during a storm, swimming through a dangerous whirlpool or cataract. A man who is timid in the water will never reign over the undines; one who is afraid of fire will never command salamanders; so long as we are liable to giddiness we must leave the sylphs in peace and forbear from irritating gnomes; for inferior spirits will only obey a power which has overcome them in their own element. When this incontestable faculty has been acquired by exercise and daring, the word of our will must be imposed on the elements by special consecrations of air, fire, water and earth. This is the indispensable preliminary of all magical operations. The air is exorcised by breathing towards the four cardinal points and saying:

The Spirit of God moved upon the waters and breathed into the face of man the breath of life. Be Michael, my leader, and Sabtabiel, my servant, in and by the light. May my breath become a word, and I will rule the spirits of this creature of air; I will curb the steeds of the sun by the will of my heart, and by the thought of my mind, and by the apple of the right eye. Therefore I do exorcise thee, creature of air, by Pentagrammaton, and in the name Tetragrammaton, wherein are firm will and true faith. Amen. *Sela: Fiat.* So be it.

The Prayer of the Sylphs must be recited next, after tracing their sign in the air with the quill of an eagle.

Prayer of the Sylphs.

Spirit of Light, Spirit of Wisdom, whose breath gives and takes away the form of all things; Thou before whom the life of every being is a shadow which transforms and a vapor which passes away; Thou who ascendest upon the clouds and dost fly upon the wings of the wind; Thou who breathest out and the limitless immensities are peopled; Thou who breathest in and all which came forth from Thee unto Thee returneth; endless movement in the eternal stability, be Thou blessed forever! We praise Thee and we bless Thee in the fleeting empire of created light, of shadows, reflections, and images; and we aspire without ceasing towards Thine immutable and imperishable splendor. May the ray of Thine intelligence and the warmth of Thy love descend on us; then what is volatile shall be fixed, the shadow shall become body, the spirit of the air shall receive a soul, and the dream be a thought. We shall be swept away no more before the tempest, but shall bridle the winged steeds of the morning, and guide the course of the evening winds, that we may flee into Thy presence. O Spirit of Spirits, O eternal Soul of Souls, O imperishable Breath of Life, O Creative Sigh, O Mouth which dost breathe forth and withdraw the life of all beings in the ebb and flow of Thine eternal speech, which is the divine ocean of movement and of truth! Amen.

Water is exorcised by imposition of hands, breathing, and speech; consecrated salt, and a little of the ash which remains in the cone of incense, are also mingled with it. The aspergillus is formed of twigs of vervain, periwinkle, sage, mint, ash, and basil, tied by a thread taken from a virgin's distaff, and provided with a handle of hazelwood from a tree which has not yet fruited; the characters of the seven spirits must be graven thereon with the magic bodkin. The salt and ash must be separately consecrated, saying:

Over the Salt.

May wisdom abide in this salt, and may it preserve our minds and bodies from all corruption, by Hochmaël, and in the virtue of Ruach-Hochmaël! May the phantoms of Hyle depart here from, that it may become a heavenly salt, salt of the earth and earth of salt, that it may feed the threshing ox, and strengthen our hope with the horns of the flying bull! Amen.

Over the Ash.

May this ash return unto the fount of living waters, may it become a fertile earth, may it bring forth the tree of life, by the Three Names, which are Netsah, Hod, and Jesod, in the beginning and in the end, by Alpha and Omega, which are in the spirit of AZOTH! Amen.

Mingling the Water, Salt and Ash.

In the salt of eternal wisdom, in the water of regeneration, and in the ash whence the new earth springeth, be all things accomplished by Eloïm, Gabriel, Raphael and Uriel, through the ages and æons! Amen.

Exorcism of the Water.

Let there be a firmament in the midst of the waters, and let it divide the waters from the waters; the things which are above are like unto things which are below, and things below are like unto things above, for the performance of the wonders of one thing. The sun is its father, the moon its mother, the wind hath carried it in the belly thereof; it ascendeth from earth to heaven, and again it descendeth from heaven to earth. I exorcise thee, creature of water, that thou mayest become unto men a mirror of the living God in His works, a fount of life and ablution of sins.

Prayer of the Undines.

Dread King of the Sea, who hast the keys of the floodgates of heaven, and dost confine the waters of the underworld in the caverns of earth; King of the deluge and the floods of the springtime; Thou who dost unseal the sources of rivers and fountains; Thou Who dost ordain moisture, which is like the blood of earth, to become the sap of plants: Thee We adore and Thee the invoke! Speak unto us, Thine inconstant and unstable creatures, in the great tumults of the sea, and we shall tremble before Thee; speak unto us also in the murmur of limpid waters, and we shall yearn for Thy love! O Immensity into which flow all the rivers of life, to be continually reborn in Thee! O ocean of infinite perfections! Height which reflects Thee in the depth, depth which exhales Thee to the height, lead us unto true life by intelligence and love! Lead us to immortality by sacrifice, that we may be found worthy one day to offer Thee water, blood and tears, for the remission of sins! Amen.

Fire is exorcised by the sprinkling of salt, incense, white resin, camphor, and sulphur, by thrice pronouncing the three names of the genii of fire: MICHAEL, king of the sun and lightning; SAMAEL, king of volcanoes; and ANAEL, prince of the astral light; and, finally, by reciting the

Prayer of the Salamanders.

Immortal, eternal, ineffable, and uncreated Father of all things, who art borne upon the ever-rolling chariot of worlds which revolve unceasingly; Lord of the ethereal immensities, where the throne of Thy power is exalted, from which height Thy terrible eyes Discern all things, and Thy holy and beautiful ears unto all things hearken, hear Thou Thy children, whom Thou didst love before the ages began; for Thy golden, Thy grand, Thine eternal majesty shines above the world and the heaven of stars! Thou art exalted over them, O glittering fire! There dost thou shine, there dost Thou commune with Thyself in Thine own splendor, and inexhaustible streams of light pour from Thine essence for the nourishment of Thine infinite spirit, which itself doth nourish all things, and forms that inexhaustible treasure of substance ever ready for generation, which adapts it and appropriates the forms Thou hast impressed on it from the beginning! From this spirit the three most holy kings Who surround Thy throne and constitute Thy court, derive also their origin, O universal Father! O sole and only Father of blessed mortals and immortals! In particular Thou hast created powers which are marvelously like unto Thine eternal thought and Thine adorable essence; Thou hast established them higher than the angels, who proclaim Thy will to the world; finally, Thou hast created us third in rank within our elementary empire. There our unceasing exercise is to praise Thee and adore Thy good pleasure; there we burn continually in our aspiration to possess Thee. O Father! O Mother, most tender of all mothers! O admirable archetype of maternity and of pure love! O Son, flower of sons! O form of all forms, soul, spirit, harmony, and number of all things! Amen.

The earth is exorcised by aspersion of water, by breathing, and by fire, with the perfumes proper for each day, and the

Prayer of the Gnomes.

King invisible, who, taking the earth as a support, didst furrow the abysses to fill them with Thine omnipotence; Thou whose name doth shake the vaults of the world, Thou who causest the seven metals to flow through the veins of the rock, monarch of the seven lights, rewarder of the subterranean toilers, lead us unto the desirable air, and to the realm of splendor. We watch and we work unremittingly, we seek and we hope, by the twelve stones of the Holy City, by the hidden talismans, by the pole of loadstone which passes through the center of the world! Savior, Savior, Savior, have pity on those who suffer, expand our hearts, detach and elevate our minds, enlarge our entire being! O stability and motion! O day clothed with night! O darkness veiled by light! O master who never keepest back the wages of Thy laborers! O silver whiteness! O golden splendor! O crown of living and melodious diamonds! Thou Who wearest the heaven on Thy finger like a sapphire ring, Thou Who concealest under earth, in the stone kingdom, the marvelous seed of stars, live, reign, be the eternal dispenser of the wealth whereof Thou hast made us the wardens! Amen.

It must be borne in mind that the special kingdom of the gnomes is at the north, that of salamanders at the south, that of sylphs at the east, and that of undines at the west. These beings influence the four temperaments of man, that is to say, the gnomes affect the melancholy, salamanders the sanguine, undines the phlegmatic and sylphs the bilious. Their signs are – the hieroglyphs of the bull for the gnomes, who are commanded with the sword; those of the lion for the salamanders, who are commanded with the bifurcated rod or the magic trident; those of the eagle for the sylphs, who are commanded by the holy pantacles; finally, those of the water-carrier for the undines, who are commanded by the cup of libations. Their respective sovereigns are Gob for the gnomes, Djîn for the salamanders, Paralda for the sylphs and Nicksa for the undines.

When an elementary spirit torments, or, at least, vexes, the inhabitants of this world, it must be conjured by air, water, fire and earth, by breathing, sprinkling, burning of perfumes, and by tracing on the earth the star of Solomon and the sacred pentagram. These figures must be perfectly correct, and drawn either with the charcoal of consecrated fire, or with a reed dipped in various colors, mixed with powdered loadstone. Then, holding the pantacle of Solomon in one hand and taking up successively the sword, rod, and cup, the conjuration of the four should be recited with a loud voice, after the following manner: –

Caput mortuum, the Lord command thee by the living and votive serpent! Cherub, the Lord command thee by Adam Jotchabah! Wandering Eagle, the Lord command thee by the wings of the Bull! Serpent, the Lord Tetragrammaton command thee by the angel and the lion! Michael, Gabriel, Raphael, and Anael! FLOW, MOISTURE, by the spirit of ELOÏM. EARTH, be established by ADAM JOTCHABAH. Spread, FIRMAMENT, by JAHUBEHU, ZEBAOTH. Fulfil, JUDGMENT, by fire in the virtue of MICHAEL. Angel of the blind eyes, obey, or pass away with this holy water! Work, winged bull, or revert to the earth, unless thou wilt that I should pierce thee with this sword! Chained eagle, obey my sign, or fly before this breathing! Writhing serpent, crawl at my feet, or be tortured by the sacred fire and give way before the perfumes that I burn in it! Water return to water; fire, burn; air, circulate; earth, revert to earth, by virtue of the pentagram, which is the morning star, and by the name of the Tetragram, which is written in the center of the cross of light! Amen.

The sign of the cross adopted by Christians does not belong to them exclusively. It is also kabbalistic, and represents the oppositions and tetradic equilibrium of the elements. We see by the occult versicle of the Lord's Prayer, which we have cited in our Doctrine, that it was originally made after two manners, or at least that it was characterized by two entirely different formulæ, one reserved for priests and initiates, the other imparted to neophytes and the profane. For example, the initiate said, raising his hand to his forehead, "For thine," then added "is," and continuing as he brought down his hand to his breast, "the kingdom," then to the left shoulder, "the justice," afterwards to the right shoulder, "and the mercy" – then clasping his hands, he added, "in the generating ages." *Tibi sunt Malkuth et Geburah et Chesed per æonas* – a sign of the cross which is absolutely and magnificently kabbalistic, which the profanations of Gnosticism have completely lost to the official and militant Church. This sign, made after this manner, should precede and terminate the conjuration of the four.

To overcome and subjugate the elementary spirits, we must never yield to their characteristic defects. Thus, a shallow and capricious mind will never rule the sylphs; an irresolute, cold, and fickle nature will never master the undines; passion irritates the salamanders; and avaricious greed makes its slaves the sport of gnomes. But we must be prompt and active, like the sylphs; pliant and attentive to images, like the undines; energetic and strong, like the salamanders; laborious and patient like the gnomes; in a word, we must overcome them in their strength without ever being overcome by their weaknesses. Once we are well established in this disposition, the whole world will be at the service of the wise operator. He will pass through the storm, and the rain will not moisten his head; the wind will not move even a fold of his garments; he will go through fire and not be burned; he will walk upon water, and will behold diamonds within the crust of the earth. These promises may appear hyperbolic, but only to vulgar understanding, for if the sage do not materially and actually perform such things, he accomplishes others which are much greater and more admirable. At the same time, it is indubitable that we may direct the elements by our will up to a certain point, and can really change or hinder their effects. For example, if it be established that persons in an ecstatic state lose their weight for the time being, why should it be impossible to walk upon the water? The convulsionaries of Saint Médard felt neither fire nor steel, and begged for the most violent blows and incredible tortures as a relief. The extraordinary climbing and miraculous equilibrium of some somnambulists are a revelation of these concealed forces of nature. But we live in a century when no one has the courage to confess the wonders that he has witnessed, and did anyone say: "I have myself beheld or performed the things which I am describing," he would be answered: "You are amusing yourself at our expense, or, otherwise, you are ill." It is far better to be silent and to act.

The metals which correspond to the four elementary forms are gold and silver for the air, mercury for water, iron and copper for fire, lead for earth. Talismans are composed from these, relative to the forces which they signify and to the effects which it is designed to obtain from them. Divination by the four elementary forms, respectively known as æromancy, hydromancy, pyromancy and geomancy, is performed after various manners, which all depend on the will and the translucid or imagination of the operator. In fact, the four elements are only instruments which assist second sight. Now, second sight is the faculty of seeing in the astral light, and it is as natural as the first or sensible and ordinary sight, but it can only operate by the abstraction of the senses. Somnambulists and ecstatics enjoy second sight naturally, but this sight is more lucid when the abstraction is more complete. Abstraction is produced by astral intoxication, that is, by an excess of light which completely saturates, and hence stupefies, the nervous system.

Sanguine temperaments are disposed to æromancy, the bilious to pyromancy, the phlegmatic to hydromancy and the melancholic to geomancy. Æromancy is confirmed by oneiromancy, or divination by dreams; pyromancy is supplemented by magnetism; hydromancy by crystallomancy; and geomancy by cartomancy. These are transpositions and complement of methods. But divination, however operated, is dangerous, or, to say the least, useless, for it disheartens will, as a consequence, impedes liberty, and tires the nervous system.

Chapter V: The Blazing Pentagram

WE proceed to the explanation and consecration of the sacred and mysterious pentagram. At this point, let the ignorant and superstitious close the book; they will either see nothing but darkness, or they will be scandalized. The pentagram, which in Gnostic schools is called the blazing star, is the sign of intellectual omnipotence and autocracy. It is the star of the magi; it is the sign of the Word made flesh; and, according to the direction of its points, this absolute magical symbol represents order or confusion, the divine lamb of Ormuz and St. John, or the accursed goat of Mendes. It is initiation or profanation; it is Lucifer or Vesper, the star of the morning or the evening. It is Mary or Lilith, victory or death, day or night. The pentagram with two points in the ascendant represents Satan as the goat of the Sabbath; when one point is in the ascendant, it is the sign of the Savior. The pentagram is the figure of the human body, having the four limbs and a single point representing the head. A human figure, head downwards, naturally represents a demon; that is, intellectual subversion, disorder, or madness. Now, if magic be a reality, if occult science be really the true law of the three worlds, this absolute sign, this sign ancient as history, and more ancient, should and does actually exercise an incalculable influence upon spirits set free from their material envelope.

The sign of the pentagram is called also the sign of the microcosm, and it represents what the Kabbalists of the book of Zohar term the microprosopus. The complete comprehension of the pentagram is the key of the two worlds. It is the absolute philosophy and natural science. The sign of the pentagram should be composed of the seven metals, or at least traced in pure gold upon white marble. It may be also drawn with vermilion upon an unblemished lambskin – the symbol of integrity and light. The marble should be virgin, that is, should never have been used for another purpose; the lambskin should be prepared under the auspices of the sun. The lamb must have been slain at Paschal time, with a new knife, and the skin must be salted with salt consecrated by magical operations. The omission of even one of these difficult and apparently arbitrary ceremonies makes void the entire success of the great works of science.

The pentagram is consecrated with the four elements; the magical figure is breathed on five times; it is sprinkled with consecrated water; it is dried by the smoke of five perfumes, namely, incense, myrrh, aloes, sulphur and camphor, to which a little white resin and ambergris may be added. The five breathings are accompanied by the utterance of names attributed to the five genii, who are Gabriel, Raphael, Anael, Samael and Oriphiel; afterwards the pentacle is placed successively at the north, south, east, west, and center of the astronomical cross, pronouncing at the same time, one after another, the consonants of the sacred tetragram, and then, in an undertone, the blessed letters Aleph and the mysterious Thau, united in the kabbalistic name of AZOTH.

The pentagram should be placed upon the altar of perfumes, and under the tripod of evocations. The operator should also wear the sign as well as that of the macrocosm, which is composed of two crossed and superposed triangles. When a spirit of light is evoked, the head of the star – that is, one of its points – should be directed towards the tripod of evocations, and the two inferior points towards the altar of perfumes. In the case of a spirit of darkness, the opposite course is pursued, but then the operator must be careful to set the end of the rod or the point of the sword upon the head of the pentagram. We have said that signs are the active voice of the word of will. Now, the word of will must be given in its completeness, so that it may be transformed into action; and a single negligence, representing an idle speech or a doubt, falsifies and paralyses the whole process, turning back upon the operator all the forces thus expended in vain. We must, therefore, absolutely abstain from magical ceremonies or scrupulously and exactly fulfil them all.

The pentagram, engraved in luminous lines upon glass by the electrical machine, also exercises a great influence upon spirits and terrifies phantoms. The old magicians traced the sign of the pentagram upon their doorsteps, to prevent evil spirits from entering and good spirits from departing. This constraint followed from the direction of the points of the star. Two points on the outer side drove away the evil; two points on the inner side imprisoned them; one only on the inner side held good spirits captive. All these magical theories, based upon the one dogma of Hermes and on the analogical deductions of science, have been confirmed invariably by the visions of ecstatics and the convulsions of cataleptics saying that they are possessed with spirits. The G which Freemasons place in the middle of the Blazing Star signifies GNOSIS and GENERATION, the two sacred words of the ancient Kabbalah. It signifies also GRAND ARCHITECT, for the pentagram on every side represents an A. By placing it in such a manner that two of its points are in the ascendant and one is below, we may see the horns, ears and beard of the hierarchic goat of Mendes, when it becomes the sign of infernal evocations.

The allegorical star of the magi is no other than the mysterious pentagram; and those three kings, sons of Zoroaster, conducted by the blazing star to the cradle of the microcosmic God, are enough in themselves to demonstration the wholly kabbalistic and truly magical beginnings of Christian doctrine. One of these kings is white, another black, and the third brown. The white king offers gold, symbol of light and life; the black king presents myrrh, image of death and of darkness; the brown king sacrifices incense, emblem of the conciliating doctrine of the two principles. Then they return into their own land by another road, to show that a new cultus is only a new path, conducting man to the one religion, that of the sacred triad and the radiant pentagram, the sole eternal *Catholicism*. St. John, in the Apocalypse, beholds this same star fall from heaven to earth. It is then called absinth or wormwood, and all the waters of the sea become bitter – striking image of the materialization of dogma, which produces fanaticism and the acridities of controversy. Then unto Christianity itself may be applied those words of Isaiah: "How has thou fallen from heaven, bright star, which wast so splendid in thy prime!" But the pentagram, profaned by men, burns ever unclouded in the right hand of the Word of Truth, and the inspired voice promises to him that overcomes the possession of the morning star – solemn restitution held out to the Star of Lucifer.

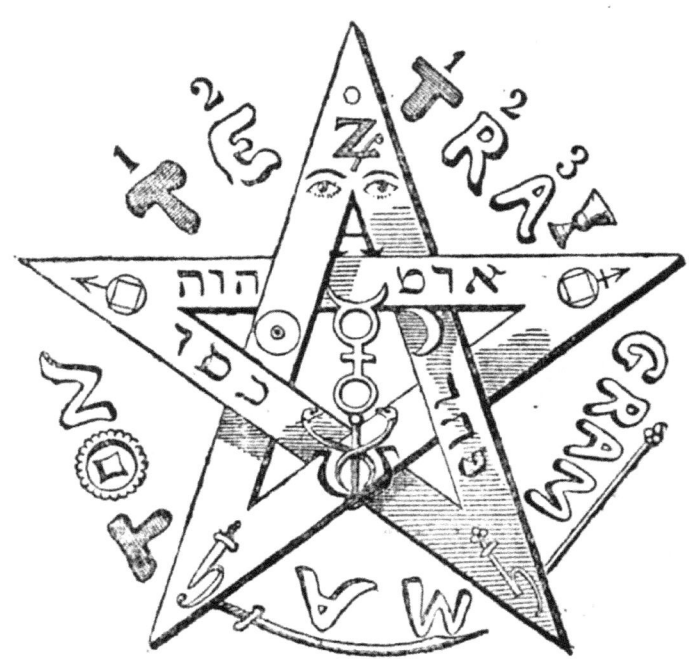

As will be seen, all mysteries of magic, all symbols of the gnosis, all figures of occultism, all kabbalistic keys of prophecy, are summed up in the sign of the pentagram, which Paracelsus proclaims to be the greatest and most potent of all signs. Is there any cause now for astonishment at the conviction of the magus as to the real influence exercised by this sign over the spirits of all hierarchies? Those who defy the sign of the cross tremble before the star of the microcosm. On the contrary, when he is conscious of failing will, the magus turns his eyes towards this symbol, takes it in his right hand and feels armed with intellectual omnipotence, provided that he is truly a king, worthy to be conducted by the star to the cradle of divine realization; provided that he *knows, dares, wills* and *keeps silent*; provided that he is familiar with the usages of the pantacle, the cup, the wand and the sword; provided, finally, that the intrepid gaze of his soul corresponds to those two eyes which the ascending point of our pentagram ever presents open.

Chapter VI: The Medium and the Mediator

TWO things, as we have already said, are necessary for the acquisition of magical power – the emancipation of the will from all servitude, and its instruction in the art of domination. The sovereign will is represented in our symbols by the woman who crushes the serpent's head, and by the radiant angel who restrains and constrains the dragon with lance and heel. In this place let us affirm without evasions that the great magical agent – the double current of light, the living and astral fire of the earth – was represented by the serpent with the head of an ox, goat or dog, in ancient theogonies. It is the dual serpent of the caduceus, the old serpent of Genesis, but it is also the brazen serpent of Moses, twisted about the *tau*, that is, the generating lingam. It is, further, the goat of the Sabbath and the Baphomet of the Templars; it is the Hyle of the Gnostics; it is the double tail of the serpent which forms the legs of the solar cock of Abraxas. In fine, it is the devil of M. Eudes de Mirville, and is really the blind force which souls must overcome if they would be freed from the chains of earth; for, unless their will can detach them from this fatal attraction, they will be absorbed in the current by the force which produced them, and will return to the central and eternal fire. The whole magical work consists, therefore, in our liberation from the folds of the ancient serpent, then in setting foot upon its head, and leading it where we will. "I will give thee all the kingdoms of the earth, and thou shalt crouch at my feet fall down and adore me," said this serpent in the evangelical mythos. The initiate should make answer: "I will not kneel down, but thou shalt crouch at my feet; nothing shalt thou give me, but I will make use of thee, and will take that which I require, for I am thy lord and master" – a reply which, in a veiled manner, is contained in that of the Savior.

We have said that the devil is not a person. It is a misdirected force, as the name indicates. An odic or magnetic current, formed by a chain of perverse wills, constitutes this evil spirit, which the Gospel calls *legion*, and this it is which precipitated the swine into the sea – another allegory of the attraction exercised on beings of inferior instincts by blind forces that can be put in operation by error and evil will. This symbol may be compared with that of the comrades of Ulysses transformed into swine by the sorceress Circe. Remark what was done by Ulysses to preserve himself and deliver his associates: he refused the cup of the enchantress, and commanded her with the sword. Circe is nature, with all her delights and allurements – to enjoy her we must overcome her. Such is the significance of the Homeric fable, for the poems of Homer, those true sacred books of ancient Hellas, contain all the mysteries of high oriental initiation.

The natural medium is, therefore, the serpent, ever active and ever seducing, of idle wills, which we must continually withstand by subjugation. Amorous, gluttonous, passionate, or idle magicians are impossible monstrosities. The magus thinks and wills; he loves nothing with desire; he rejects nothing in rage. The word *passion* signifies a passive state, and the magus is invariably active, invariably victorious. The attainment of this realization is the crucial difficulty of the transcendent sciences; so, when the magus accomplishes his own creation, the great work is fulfilled, at least as concerns cause and instrument. The great agent or natural mediator of human omnipotence cannot be overcome or directed save by an *extra-natural* mediator, which is an emancipated will. Archimedes postulated a fulcrum outside the world in order to raise the world. The fulcrum of the magus is the intellectual cubic stone, the philosophical stone of AZOTH – that is, the doctrine of absolute reason and universal harmonies by the sympathy of contraries.

One of our most fertile writers, and one of those who are least fixed in their ideas, M. Eugène Sue, has founded a vast romance-epic upon an individuality whom he strives to render odious, who becomes interesting against the will of the novelist, so abundantly does he gift him with patience, audacity, intelligence and genius. We are in the presence of a kind of Sixtus V. – poor, temperate, passionless, holding the entire world entangled in the web of his skillful combinations. This man excites at will the passions of his enemies, destroys them by means of one another, invariably reaches the point he has kept in view, and this without noise, without ostentation, and without imposture. His object is to free the world of a society which the author of the book believes to be dangerous and malignant, and to attain it no cost is too great; he is ill lodged, ill clothed, nourished like the refuse of humanity, but ever fixed upon his work. Consistently with his intention, the author depicts him as wretched, filthy, hideous, repulsive to the touch, and horrible to the sight. But supposing this very exterior is a means of disguising the enterprise, and so of more surely attaining it, is it not proof positive of sublime courage? When Rodin becomes pope, do you think that he will still be ill clothed and dirty? Hence M. Eugene Sue has missed his point; his object was to deride superstition and fanaticism, but that which he attacks is intelligence, strength, genius, the most signal human virtues. Were there many Rodins among the Jesuits, were there one even, I would not give much for the success of the opposite party, in spite of the brilliant and maladroit special pleadings of its illustrious advocates.

To will well, to will long, to will always, but never to lust after anything, such is the secret of power, and this is the magical arcanum which Tasso brings forward in the persons of the two knights who come to deliver Rinaldo and to destroy the enchantments of Armida. They withstand equally the most charming nymphs and most terrible wild beasts. They remain without desires and without fear, and hence they attain their end. Does it follow from this that a true magician inspires more fear than love? I do not deny it, and while abundantly recognizing how sweet are the allurements of life, while doing full justice to the gracious genius of Anacreon, and to all the youthful efflorescence of the poetry of love, I invite the estimable votaries of pleasure to regard the transcendental sciences merely as a matter of curiosity, and never to approach the magical tripod; the great works of science are deadly for pleasure.

The man who has escaped from the chain of instincts will first of all realize his omnipotence by the submissiveness of animals. The history of Daniel in the lions' den is no fable, and more than once, during the persecutions of infant Christianity, this phenomenon recurred in the presence of the whole Roman people. A man seldom has anything to fear from an animal of which he is not afraid. The bullets of Jules

Gérard, the lion-killer, are magical and intelligent. Once only did he run a real danger; he allowed a timid companion to accompany him, and, looking upon this imprudent person as lost beforehand, he also was afraid, not for himself but for his comrade. Many persons will say that it is difficult and even impossible to attain such resolution, that strength in volition and energy in character are natural gifts. I do not dispute it, but I would point out also that habit can reform nature; volition can be perfected by education, and, as I have said before, all magical, like all religious, ceremonial has no other end but thus to test, exercise, and habituate the will by perseverance and by force. The more difficult and laborious the exercises, the greater their effect, as we have now advanced far enough to see.

If it has been hitherto impossible to direct the phenomena of magnetism, it is because an initiated and truly emancipated operator has not yet appeared. Who can boast that he is such? Have we not ever new self-conquests to make? At the same time, it is certain that nature will obey the sign and word of one who feels himself strong enough to be convinced of it. I say that nature will obey; I do not say that she will belie herself or disturb the order of her possibilities. The healing of nervous diseases by word, breath, or contact; resurrection in certain cases; resistance of evil wills sufficient to disarm and confound murderers; even the faculty of making one's self invisible by troubling the sight of those whom it is important to elude; all this is a natural effect of projecting or withdrawing the astral light. Thus, was Valentius dazzled and terror-struck on entering the temple of Cesarea, even as Heliodorus of old, overcome by a sudden madness in the temple of Jerusalem, believed himself scourged and trampled by angels. Thus, also the Admiral de Coligny imposed respect on his assassins, and could only be dispatched by a madman who fell upon him with averted face. What rendered Joan of Arc invariably victorious was the fascination of her faith and the miracle of her audacity; she paralyzed the arms of those who would have assailed her, and the English may have very well been sincere in regarding her as a witch or a sorceress. As a fact, she was a sorceress unconsciously, herself believing that she acted supernaturally, while she was really disposing of an occult force which is universal and invariably governed by the same laws.

The magus-magnetizer should have command of the natural medium, and consequently of that astral body by which our soul communicates with our organs. He can say to the material body, "Sleep!" and to the sidereal body, "Dream!" Thereupon, the aspect of visible things changes, as in hashish-visions. Cagliostro is said to have possessed this power, and he increased its action by means of fumigations and perfumes; but true magnetic ability should transcend these auxiliaries, all more or less inimical to reason and destructive of health. M. Ragon, in his learned work on Occult Masonry, gives the recipe for a series of medicaments suitable for the exaltation of somnambulism. It is by no means a knowledge to be despised, but prudent magists should avoid its practice.

The astral light is projected by glance, by voice, and by the thumb and palm of the hand. Music is a potent auxiliary of the voice, and hence comes the word *enchantment*. No musical instrument has more enchantment than the human voice, but the far-away notes of a violin or harmonica may augment its power. The subject whom it is proposed to overcome is in this way prepared; and when he is half-deadened and, as it were, enveloped by the charm, the hands should be extended towards him; he should be commanded to sleep or to see, and he will obey despite himself. Should he resist, a fixed glance must be directed towards him, one thumb must be placed between his eyes and the other on his breast, touching him lightly with a single and swift contact; the breath must be slowly drawn in and again breathed gently and warmly forth, repeating in a low voice, "Sleep!" or "See!"

Chapter VII: The Septenary of Talismans

Ceremonies, vestments, perfumes, characters and figures, being, as we have stated, necessary to enlist the imagination in the education of the will, the success of magical works depends upon the faithful observation of all the rites, which are in no sense fantastic or arbitrary, having been transmitted to us by antiquity, and permanently subsisting by the essential laws of analogical realization and of the correspondence which invariably connects ideas and forms. Having spent many years in consulting and comparing all the most authentic grimoires and magical rituals, we have succeeded, not without labor, in reconstituting the ceremonial of universal and primeval magic. The only serious books which we have seen upon this subject are in manuscript, written in conventional characters which we have deciphered by the help of the polygraphy of Trithemius. The importance of others consists wholly in the hieroglyphs and symbols which adorn them, the truth of the images being disguised under the superstitious fictions of a mystifying text. Such, for example, is the "Enchiridion" of Pope Leo III, which has never been printed with its true figures, and we have reconstructed it for our own use after an ancient manuscript. The rituals known under the name of "Clavicles of Solomon" are very numerous. Many have been printed, while others remain in manuscripts, transcribed with great care. An exceedingly fine and elegantly written example is preserved in the Imperial Library; it is enriched with pantacles and characters most of which have been reproduced in the magical calendars of Tycho-Brahe and Duchentau. Lastly, there are printed clavicles and grimoires which are catch-penny mystifications and impostures of dishonest publishers. The book so notorious and decried formerly under the name of "Little Albert" belongs mainly to the latter category; some talismanic figures, and some calculations borrowed from Paracelsus, are its only serious parts.

In any matter of realization and ritual, Paracelsus is an imposing magical authority. No one has accomplished works greater than his, and for that very reason he conceals the virtue of ceremonies and merely teaches in his occult philosophy the existence of the magnetic agent of omnipotence of will; he also sums the whole science of characters in two signs, the macrocosmic and microcosmic stars. It was sufficient for the adepts, and it was important not to initiate the vulgar. Paracelsus therefore did not teach the ritual, but he practiced, and his practice was a sequence of miracles.

We have spoken of the magical importance of the triad and tetrad. Their combination constitutes the great religious and kabbalistic number which represents the universal synthesis and comprises the sacred septenary. In the belief of the ancients, the world is governed by seven secondary causes – *secundœi,* as Trithemius calls them – which are the universal forces designated by Moses under the plural name of Eloïm, gods. These forces, analogous and contrary to one another, produce equilibrium by their contrasts, and rule the motion of the spheres. The Hebrews termed them the seven great archangels, giving them the names of Michael, Gabriel, Raphael, Anael, Samael, Zadkiel, and Oriphiel. The Christian Gnostics named the four last Uriel, Barachiel, Sealtiel, and Jehudiel. Other nations attributed to these spirits the government of the seven chief planets, and assigned to them the names of their chief divinities. All believed in their relative influence; astronomy divided the antique heaven between them and allotted the seven days of the week to their successive government. Such is the reason of the various ceremonies of the magical week and the septenary cultus of the planets. We have already observed that here the planets are signs and nothing else; they have the influence which universal faith attributes because they are more truly the stars of the human mind than the orbs of heaven. The sun, which antique magic always regarded as fixed, could be only a planet for the vulgar; hence it represents the day of repose in the week, which we term Sunday without knowing why, the day of the sun among the ancients.

The seven magical planets correspond to the seven colors of the prism and the seven notes of the musical octave; they represent also the seven virtues, and, by opposition, the seven vices of Christian ethics. The seven sacraments correspond equally to this great universal septenary. Baptism, which consecrates the element of water, corresponds to the Moon; ascetic penance is under the auspices of Samael, the angel of Mars; confirmation, which imparts the spirit of understanding and communicates to the true believer the gift of tongues, is under the auspices of Raphael, the angel of Mercury; the Eucharist substitutes sacramental realization of God made man for the empire of Jupiter; marriage is consecrated by the angel Anael, the purifying genius of Venus; extreme unction is the safeguard of the sick about to fall under the scythe of Saturn; and orders, consecrating the priesthood of light, is marked, more especially by the characters of the

Sun. Almost all these analogies were observed by the learned Dupuis, who thence concluded that all religions were false, instead of recognizing the sanctity and perpetuity of a single dogma, ever reproduced in the universal symbolism of successive religious forms. He failed to understand the permanent revelation transmitted to human genius by the harmonies of nature, and beheld only a catalogue of errors in that chain of speaking images and eternal truths.

Magical works are also seven in number:

1. works of light and riches, under the auspices of the Sun;
2. works of divination and mystery, under the invocation of the Moon;
3. works of skill, science and eloquence, under the protection of Mercury;
4. works of wrath and chastisement, consecrated to Mars;
5. works of love, favored by Venus;
6. works of ambition and intrigue, under the influence of Jupiter;
7. works of malediction and death, under the patronage of Saturn.

In theological symbolism, the Sun represents the word of truth; the Moon, religion itself; Mercury, the interpretation and science of mysteries; Mars, justice; Venus, mercy and love; Jupiter, the risen and glorious Savior; Saturn, God the Father, or the Jehovah of Moses. In the human body, the Sun is analogous to the heart, the Moon to the brain, Jupiter to the right hand, Saturn to the left, Mars to the left foot, Venus to the right, Mercury to the generative organs, whence an androgyne figure is sometimes attributed to this planet. In the human face, the Sun governs the forehead, Jupiter the right and Saturn the left eye; the Moon rules between both at the root of the nose, the two phlanges of which are governed by Mars and Venus; finally, the influence of Mercury is exercised over mouth and chin. Among the ancients these notions constituted the occult science of physiognomy, afterwards recovered imperfectly by Lavater.

The magus who intends undertaking the works of light must operate on a Sunday, from midnight to eight in the morning, or from three in the afternoon to ten in the evening. He should wear a purple vestment, with tiara and bracelets of gold. The altar of perfumes and the tripod of sacred fire must be encircled by wreaths of laurel, heliotrope and sunflowers; the perfumes are cinnamon, strong incense, saffron and red sandal; the ring must be of gold, with a chrysolith or ruby; the carpet must be of lion-skins, the fans of sparrow-hawk feathers. On Monday the robe is white, embroidered with silver, and having a triple collar of pearls, crystals and selenite; the tiara must be covered with yellow silk, emblazoned with silver characters forming the Hebrew monogram of Gabriel, as given in the "Occult Philosophy" of Agrippa; the perfumes are white sandal, camphor, amber, aloes and pulverized seed of cucumber; the wreaths are mugwort, moonwort and yellow ranunculuses. Tapestries, garments and objects of a black color must be avoided; and no metal except silver should be worn on the person. On Tuesday, a day for the operations of vengeance, the color of the vestment should be that of flame, rust, or blood, with belt and bracelets of steel. The tiara must be bound with gold; the rod must not be used, but only the magical dagger and sword; the wreaths must be of absinthe and rue, the ring of steel, with an amethyst for precious stone. On Wednesday, a day favorable for transcendent science, the vestment should be green, or shot with various colors; the necklace of pearls in hollow glass beads containing mercury; the perfumes benzoin, mace, and storax; the flowers, narcissus, lily, herb mercury, fumitory, and marjorlane; the jewel should be the agate. On Thursday, a day of great religious and political operations, the vestment should be scarlet and, on the forehead, should be worn a brass tablet, with the character of the spirit of Jupiter and the three words: GIARAR, BETHOR, SAMGABIEL; the perfumes are incense, ambergris, balm, grain of paradise, macis, and saffron; the ring must be enriched with an emerald or sapphire; the wreaths and crowns should be oak, poplar, fig and pomegranate leaves. On Friday, the day for amorous operations, the vestment should be of sky blue, the hangings of green and rose, the ornaments of polished copper, the crowns of violets, the wreaths of roses, myrtle, and olive; the ring should be enriched with a turquoise; lapis-lazuli and beryl will answer for tiara and clasps; the fans must be of swan's feathers; and the operator must wear upon his breast a copper talisman with the character of Anael and the words: AVEEVA VADELILITH. On Saturday, a day of funereal operations, the vestment must be black or brown, with characters embroidered in black or orange colored silk; on the neck must be worn a leaden medal with the character of Saturn and the words: ALMALEC, APHIEL, ZARAHIEL; the perfumes should be diagridrium, scammony, alum, sulphur and assafœtida; the ring should be adorned with an onyx; the garlands should be of ash, cypress and hellebore; on the onyx of the ring, during the hours of Saturn, the double head of Janus should be engraved with the consecrated awl.

Such are the antique magnificences of the secret cultus of the magi. With similar appointments the great magicians of the Middle Ages proceeded to the daily consecration of talismans corresponding to the seven genii. We have already said that a pantacle is a synthetic character resuming the entire magical doctrine in one of its special conceptions. It is, therefore, the full expression of a completed thought and will; it is the signature of a spirit. The ceremonial consecration of this sign attaches to it still more strongly the intention of the operator and establishes a veritable magnetic chain between himself and the pantacle. Pantacles may be indifferently traced upon virgin parchment, paper, or metals. What is termed a talisman is a sheet of metal, bearing either pantacles or characters, and having received a special consecration for a defined intention. In a learned work on magical antiquities, Gaffarel has scientifically demonstrated the real power of talismans, and the confidence in their virtue is otherwise so strong in nature that we gladly bear about us some memorial of those we love, keepsakes of those we love, persuaded that such keepsakes will preserve us from danger and increase our happiness. Talismans are made of the seven Kabbalistic metals, and, when the days and hours are favorable, the required and determined signs are engraved thereon. The figures of the seven planets, with their magical squares following Paracelsus, are found in the "Little Albert." It should be observed that Paracelsus replaces the figure of Jupiter by that of a priest, a substitution not wanting in a well-defined mysterious intention. But the allegorical and mythological figures of the seven spirits have now become too classical and too vulgar to be any longer successfully engraved on talismans; we must recur to more learned and expressive signs. The pentagram should be invariably engraved upon one side of the talisman, with a circle for the Sun, a crescent for the Moon, for Mars a sword, a G for Venus, for Jupiter a crown and a scythe for Saturn. The reverse must bear the sign of Solomon, that is, the six-pointed star composed of two superposed triangles; in the center there is placed a human figure for the talismans of the Sun, a chalice for those of the Moon, a dog's head for those of Mercury, an eagle's for those of Jupiter, a lion's head for those of Mars, a dove's for those of Venus, and a bull's or goat's for those of Saturn. The names of the seven angels are added either in Hebrew, in Arabic, or in magical characters like those of the alphabet of Trithemius. The two triangles of Solomon may be replaced by the double cross of the wheels of Ezekiel, which is found on a great number of ancient pantacles and is, as we have observed in our Doctrine the key to the trigrammes of Fohi.

Precious stones may be also employed for amulets and talismans; but all objects of this nature, whether metals or gems, must be kept carefully in silken bags of a color analogous to that of the spirit of the planet, perfumed with the perfumes of the corresponding day, and preserved from all impure glances and contacts. Thus, pantacles and talismans of the Sun must not be seen or touched by deformed or misshapen persons, or by immoral women; those of the Moon are profaned by the looks and hands of debauched men and menstruating females; those of Mercury lose their virtue if seen or touched by paid priests; those of Mars must be concealed from cowards; those of Venus from depraved men and men under a vow of celibacy; those of Jupiter from the impious; those of Saturn from virgins and children, not that their looks or touches can ever be impure, but because the talisman would bring them misfortune and thus lose all its virtue.

Crosses of honor and other kindred decorations are veritable talismans, which increase personal value and merit; they are consecrated by solemn investiture, and public opinion can impart to them a prodigious power. Sufficient attention has not been paid to the reciprocal influence of signs on ideas and of ideas on signs; it is not less true that the revolutionary work of modern times, for example, has been resumed symbolically in its entirety by the Napoleonic substitution of the Star of Honor for the Cross of St. Louis. It is the

Pentagram in place of the Labarum; it is the reconstitution of the symbol of light; it is the Masonic resurrection of Adonhiram. They say that Napoleon believed in his star, and could he have been persuaded to explain what he meant by this star, it would have proved to be his genius; he would therefore have adopted the pentagram for his sign, that symbol of human sovereignty by intelligent initiative. The mighty soldier of the Revolution knew little, but he divined almost everything; so was he the greatest instinctive and practical magician of modern times; the world is still full of his miracles, and the country people will never believe that he is dead.

Blessed and indulgenced objects, touched by holy images or venerable persons; chaplets from Palestine; the *Agnus Dei*, composed of the wax of the Paschal candle and the annual remnants of holy chrism; scapulas and medals, are all true talismans. One such medal has become popular in our own day, and even those who are devoid of religion suspend it from the necks of their children. Moreover, its figures are so perfectly Kabbalistic that it is truly a marvelous double pantacle. On the one side is the great initiatrix, the heavenly mother of the Zohar, the Isis of Egypt, the Venus-Urania of the Platonists, the Mary of Christianity, throned upon the world and setting one foot upon the head of the magical serpent. She extends her two hands in such a manner as to form a triangle, of which her head is the apex; her hands are open and radiant, thus making a double triangle, with all the beams directed towards the earth, evidently representing the emancipation of intelligence by labor. On the other side is the double Tau of the hierophants, the Lingam with the double Cteis or the triple Phallus, supported, with interlacement and repeated insertion, by the kabbalistic and masonic M, representing the square between the two Pillars JAKIN and BOHAS; below are placed, upon the same plane, two loving and suffering hearts, encircled by twelve pentagrams. Everyone will tell you that the wearers of this medal do not attach such significance to it, but it is only on that account more absolutely magical; having a double sense and, consequently, a double virtue. The ecstatic on the authority of whose revelations this talisman was engraved, had already beheld it existing perfectly in the astral light, which demonstrates the intimate connection of ideas and signs, giving a new sanction to the symbolism of universal magic.

The greater the importance and solemnity brought to bear on the execution and consecration of talismans and pantacles, the more virtue they acquire, as will be understood upon the evidence of the principles which we have established. Such consecration should take place on the days that we have indicated, with the appointments which we have given in detail. Talismans are consecrated by the four exorcised elements, after conjuring the spirits of darkness by the Conjuration of the Four. Then, taking up the pantacle, and sprinkling it with some drops of magical water, say:

In the name of Eloim and by the spirit of the living waters, be thou unto me a sign of light and a sacrament of will!

Presenting it to the smoke of the perfumes:

By the brazen serpent which destroyed the serpents of fire, be thou, etc.

Breathing seven times upon the pantacle or talisman: By the firmament and spirit of the voice, be thou, etc.

Lastly, placing some particles of purified earth or salt triad-wise upon it:

In the salt of earth, and by the virtue of eternal life, be thou etc.

Then recite the Conjuration of the Seven as follows, alternately casting a pastille of the seven perfumes into the sacred fire:

In the name of Michael, may Jehovah command thee, and drive thee hence, Chavajoth!

In the name of Gabriel, may Adonai command thee, and drive thee hence, Belial!

In the name of Raphael, begone before Elchim, Sachabiel!

By Samael Zebaoth, and in the name of Eloïm Gibor, get thee hence, Adrameleck!

By Zachariel and Sachiel-Meleck, be obedient unto Elvah, Samgabiel!

By the divine and human name of Shaddaï, and by the sign of the pentagram which I hold in my right hand; in the name of the angel Anael, by the power of Adam and of Heva, who are Jotchavah, begone, Lilith! Let us rest in peace, Nahemah!

By the holy Eloïm and by the names of the genii Cashiel, Sehaltiel, Aphiel, and Zarahiel, at the command of Orifiel, depart from us, Moloch! We deny thee our children to devour.

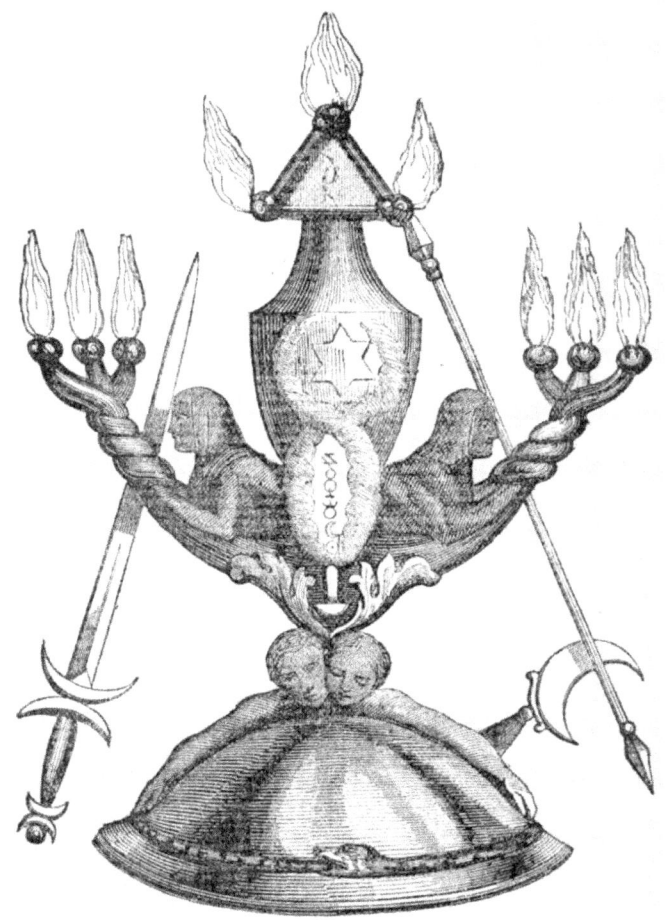

MAGICAL INSTRUMENTS.
Lamp, rod, sword, and dagger.

The most important magical instruments are the rod, the sword, the lamp, the chalice, the altar, and the tripod. In the operations of transcendent and divine magic, the lamp, rod and chalice are used; in the works of black magic, the rod is replaced by the sword and the lamp by the candle of Cardan. We shall explain this difference in the chapter devoted to black magic. Let us come now to the description and consecration of the instruments. The magical rod, which must not be confounded with the simple divining rod, with the fork of necromancers, or the trident of Paracelsus, the true and absolute magical rod must be one perfectly straight beam of almond or hazel, cut at a single blow with the magical pruning-knife or golden sickle, before the rising of the sun, at that moment when the tree is ready to blossom. It must be pierced through its whole length without splitting or breaking it, and a long needle of magnetized iron must fill its entire extent; to one of its extremities must be fitted a polyhedral prism, cut in a triangular shape, and to the other a similar figure of black resin. Two rings, one of copper, and one of zinc, must be placed at the center of the rod; subsequently, the rod must be gilt at the resin end, and silvered at the prism end as far as the ringed center; it must then be covered with silk, the extremities not included. On the copper ring these characters must be engraved: ירושליסהקדשה and on the zinc ring: המלך שלמה. The consecration of the rod must last seven days, beginning at the new moon, and should be made by an initiate possessing the great arcana, and having himself a consecrated rod. This is the transmission of the magical secret, which has never ceased since the shrouded origin of the transcendent science. The rod and the other instruments, but the rod above all, must be concealed with care, and under no pretext should the Magus permit them to be seen or touched by the profane: otherwise they will lose all their virtue. The mode of transmitting the rod is one of the arcana of science, the revelation of which is never permitted. The length of the magical rod must not exceed that of the operator's arm; the magician must never use it unless he is alone, and even then, should not touch it without necessity. Many ancient Magi made it the length of the forearm and concealed it beneath their long mantles, showing only the simple divining rod in public, or some allegorical scepter made of ivory or ebony, according to the nature of the works. Cardinal Richelieu, always a thirst for power, sought through his whole life the transmission of the rod, without being able to find it. His Kabalist Gaffarel could furnish him with sword and talismans alone; this was possibly the secret motive for the cardinal's hatred of Urban Grandier, who knew something of his weaknesses. The secret and prolonged conversations of Laubardement with the unhappy priest some hours before his final torture, and those words of a friend and confidant of the latter, as he went forth to death – "You are a clever man, monsieur, do not destroy yourself" – afford considerable food for thought.

The magical wand is the *verendum* of the magus; it must not even be mentioned in any clear and precise manner; no one should boast of its possession, nor should its consecration ever be transmitted except under the conditions of absolute discretion and confidence.

The sword is less occult, and is made in the following manner: It must be of pure steel, with a cruciform copper handle having three pommels, as represented in the enchiridion of Leo III, or with the guard of a double crescent, as in our own figure. On the middle knot of the guard, which should be covered with a golden plate, the sign of the macrocosm must be inscribed on one side and that of the microcosm on the other. The Hebrew monogram of Michael, as found in Agrippa, must be engraved on the pommel; on one side of the blade must be these characters: באילים יהוה מי כמכה, and on the other the monogram of the Labarum of Constantine, followed by the words: *Vince in hoc, Deo duce, comite ferro*. For the authenticity and exactitude of these figures, see the best ancient editions of the "Enchiridion." The consecration of the sword must take place on a Sunday, during the hours of the sun, under the invocation of Michael. The blade of the sword must be placed in a fire of laurel and cypress; it must then be dried and polished with ashes of the sacred fire, moistened with the blood of a mole or serpent, the following words being said:

> Be thou unto me as the sword of Michael; by virtue of Eloïm Sabaoth, may spirits of darkness and reptiles of earth flee away from thee!

It is then fumigated with the perfumes of the sun, and wrapped up in silk, together with branches of vervain, which should be burned on the seventh day.

The magical lamp must be composed of the four metals – gold, silver, brass and iron; the pedestal should be of iron, the mirror of brass, the reservoir of silver, the triangle at the apex of gold. It should be provided with two branches composed of a triple pipe of three intertwisted metals, in such a manner that each arm has a triple conduit for the oil; there must be nine wicks in all, three at the top, and three in each arm. The seal of Hermes must be engraved on the pedestal, over which must be the two-headed androgyne of Khunrath. A serpent devouring its own tail must encircle the lower part. The sign of Solomon must be chased on the reservoir. Two globes must be fitted to this lamp, one adorned with a transparent pictures, representing the seven genii, while the other, of larger size and duplicated, should contain variously tinted waters in four compartments. The whole instrument should be placed in a wooden pillar, revolving on its own axis, and permitting a ray of light to escape, as required, and fall on the altar smoke at the moment for the invocations. This lamp is a great aid to the intuitive operations of slow imaginations, and for the immediate creation in the presence of magnetized persons of forms alarming in their actuality, which, being multiplied by the mirrors, will magnify suddenly, and transform the operator's cabinet into a vast hall filled with visible souls; the intoxication of the perfumes and the exaltation of the invocations will change this fantasia into a real dream; persons known formerly will be recognized; phantoms will speak; and something extraordinary and unexpected will follow the closing of the light within the pillar and the increase of the fumigations.

Chapter VIII: A Warning to the Imprudent

THE operations of science are not devoid of danger, as we have stated several times. They may end in madness for those who are not established firmly on the basis of supreme, absolute and infallible reason. Terrible and incurable diseases can be occasioned by excessive nervous excitement. Swoons and death itself, as a consequence of cerebral congestion, may result from imagination when it is unduly impressed and terrified. We cannot sufficiently dissuade nervous persons, and those who are naturally disposed to exaltation, women, young people, and all who are not habituated in perfect self-control and command of fear. In the same way, there can be nothing more dangerous than to make magic a pastime, or, as some do, part of an evening's entertainment. Even magnetic experiments, performed under such conditions, can only exhaust the subjects, mislead opinions, and defeat science. The mysteries of life and death cannot be made sport of with impunity, and things which are to be taken seriously must be treated not only seriously but also with the greatest reserve. Never yield to the desire of convincing others by phenomena. The most astounding phenomena would not be proofs to those who are not already convinced. They can always be attributed to ordinary artifices and the magus included among the more or less skillful followers of Robert Houdin or Hamilton. To require prodigies as a warrant for believing in science is to show one's self unworthy or incapable of science. SANCTA SANCTIS. Contemplate the twelfth figure of the Tarot-Keys, remember the grand symbol of Prometheus, and be silent. All those magi who divulged their works died violently, and many were driven to suicide, like Cardan, Schroepffer, Cagliostro, and others. The magus should live in retirement, and be approached with difficulty. This is the significance of the ninth key of the Tarot, where the initiate appears as a hermit enveloped completely in his cloak. Such retirement must not, however, be one of isolation; attachments and friendships are necessary, but he must choose with care and preserve them at all price. He must have also another profession other than that of magician; magic is not a trade.

In order to devote ourselves to ceremonial magic, we must be free from anxious preoccupations; we must be in position to procure all the instruments of the science, and be able to make them when needed; we must also possess an inaccessible laboratory, in which there will be no danger of ever being surprised or disturbed. Then, and this is an indispensable condition, we must know how to equilibrate forces and restrain the zeal of our own initiative. This is the meaning of the eighth key of Hermes, wherein a woman is seated between two pillars, with an upright sword in one hand and a balance in the other. To equilibrate forces, they must be simultaneously maintained and caused to act alternately; the use of the balance represents this double action. The same arcanum is typified by the dual cross in the pantacles of Pythagoras and Ezekiel (see the plate which appears in chapter 18 in the "Doctrine"), where the crosses equilibrate each other and the planetary signs are always in opposition. Thus, Venus is the equilibrium of the works of Mars; Mercury moderates and fulfils the operations of the Sun and Moon; Saturn balances Jupiter. It was by means of this antagonism between the ancient gods that Prometheus, that is to say, the genius of science, contrived to enter Olympus and carry off fire from heaven. Is it necessary to speak more clearly? The milder and calmer you are, the more effective will be your anger; the more energetic you are, the more precious will be your forbearance; the more skillful you are, the better will you profit by your intelligence and even by your virtues; the more indifferent you are, the more easily will you make yourself loved. This is a matter of experience in the moral order, and is literally realized in the sphere of action. Human passions produce blindly the opposites of their unbridled desire, when they act without direction. Excessive love arouses antipathy; blind hate counteracts and scourges itself; vanity leads to abasement and the most cruel humiliations. Thus, the Great Master revealed a mystery of positive magical science when He said: "Forgive your enemies, do good to those that hate you, so shall ye heap coals of fire upon their heads." Perhaps this kind of pardon seems hypocrisy and bears a strong likeness to refined vengeance. But we must remember that the magus is sovereign, and a sovereign never avenges because he has the right to punish; in the exercise of this right he performs his duty and is implacable as justice. Let it be observed, for the rest, so that no one may misinterpret my meaning, that it is a question of chastising evil by good and opposing mildness to violence. If the exercise of virtue be a flagellation for vice, no one has the right to demand that it should be spared, or that we should take pity on its shame and its sufferings.

The man who dedicates himself to the works of science must take moderate daily exercise, abstain from prolonged vigils, and follow a wholesome and regular rule of life. He must avoid the effluvia of putrefaction, the neighborhood of stagnant water, and indigestible or impure food. Above all, he must seek daily relaxation from magical preoccupations amongst material cares, or in labor, whether artistic, industrial or commercial. The way to see well is not to be always looking; and he who spends his whole life upon one object will end without attaining it. Another precaution must be observed equally, and that is never to experiment when ill.

The ceremonies being, as we have said, artificial methods for creating a habit of will become unnecessary when the habit is confirmed. It is in this sense, and addressing himself solely to perfect adepts, that Paracelsus proscribes ceremonial work in his *Occult Philosophy*. They must be progressively simplified before they are dispensed with altogether, and in proportion to the experience we obtain in acquired powers, and established habit in the exercise of extra-natural will.

Chapter IX: The Ceremonial of Initiates

The science is preserved by silence and perpetuated by initiation. The law of silence is not, therefore, absolute and inviolable, except relatively to the uninitiated multitude. The science can be only transmitted by speech. The sages must therefore must speak occasionally. Yes, they must speak, not to disclose, but to lead others to discover. *Noli ire, fac venire*, was the device of Rabelais, who, being master of all the sciences of his time, could not be unacquainted with magic. We have, consequently, to reveal here the mysteries of initiation. The destiny of man, as we have said, is to make or create himself; he is, and he will be, the son of his works, both for time and eternity. All men are called on to compete, but the number of the elect – that is, of those who succeed – is invariably small. In other words, the men who are desirous to attain are numbered by multitudes, but the *chosen* are few. Now, the government of the world belongs by right to the flower of mankind, and when any combination or usurpation prevents their possessing it, a political or social cataclysm ensues. Men who are masters of themselves become easily masters of others; but it is possible for them to hinder one another if they disregard the laws of discipline and of the universal hierarchy. To be subject to a discipline in common, there must be a community of ideas and desires, and such a communion cannot be attained except by a common religion established on the very foundations of intelligence and reason. This religion has always existed in the world, and is that only which can be called one, infallible, indefectible, and veritably catholic – that is, universal. This religion, of which all others have been successively the veils and the shadows, is that which demonstrates being by being, truth by reason, reason by evidence and common sense. It is that which proves by realities the reasonable basis of hypotheses, and forbids reasoning upon hypotheses independently of realities. It is that which is grounded on the doctrine of universal analogies, but never confounds the things of science with those of faith. It can never be of faith that two and one make more or less than three; that in physics the contained can exceed the container; that a solid body, as such, can act like a fluidic or gaseous body; that, for example, a human body can pass through a closed door without dissolution or opening. To say that one believes such a thing is to talk like a child or a fool; yet it is no less insensate to define the unknown, and to argue from hypothesis to hypothesis, till we come to deny evidence *à priori* for the affirmation of precipitate suppositions. The wise man affirms what he knows, and believes in what he does not know only in proportion to the reasonable and known necessities of hypothesis.

But this reasonable religion is not adapted for the multitude, for which fables, mysteries, definite hopes, and terrors having a physical basis, are needful. It is for this reason that the priesthood has been established in the world. Now, the priesthood is recruited by initiation. Religious forms perish when initiation ceases in the sanctuary, whether by the betrayal of the mysteries or by their neglect and oblivion. The Gnostic disclosures, for example, alienated the Christian Church from the high truths of the Kabbalah, which contains all secrets of transcendental theology. Hence, the blind, having become leaders of the blind, great obscurities, great lapses, and deplorable scandals have followed. Subsequently, the sacred books, of which the keys are all kabbalistic, from Genesis to the Apocalypse, have become so little intelligible to Christians, that pastors have judged it necessary to forbid their being read by the uninstructed among believers. Taken literally, and understood materially, such books could be only an inconceivable tissue of absurdities and scandals, as the school of Voltaire has too well demonstrated. It is the same with all the ancient dogmas, their brilliant theogonies and poetic legends. To say that ancients of Greece believed in the love-adventures of Jupiter, or those of Egypt in the cynocephalus and sparrow-hawk, is to exhibit as much ignorance and bad faith as would be shown by maintaining that Christians adore a triple God, composed of an old man, an executed criminal, and a pigeon. The ignorance of symbols is invariably calumnious. For this reason, we should always guard against the derision of that which we do not know, when its enunciation seems to involve some absurdity or even singularity, as a course no less wanting in good sense than to admit the same without discussion and examination.

Prior to anything which may please or displease ourselves, there is a truth – that is to say, a reason – and by this reason must our actions be regulated rather than by our desires, if we would create that intelligence within us which is the *raison d'être* of immortality, and that justice which is the law thereof. A man who is truly man can will only that which he should reasonably and justly do; so, does he silence lusts and fears that he may hearken solely to reason. Now, such a man is a natural king and a voluntary priest for the wandering multitudes. Hence it was that the end of the old initiations was indifferently termed the sacerdotal art and the royal art. The antique magical associations were seminaries for priests and kings, and admission could only be obtained by truly sacerdotal and royal works; that is, by transcending all weakness of nature. We will not repeat here what is found everywhere concerning Egyptian initiations, perpetuated, but with diminished power, in the secret societies of the Middle Ages. Christian radicalism, founded upon a false understanding of the words: "Ye have one father, one master, and ye are all brethren," dealt a terrible blow at the sacred hierarchy. Since that time, sacerdotal dignities have become a matter of intrigue or of chance; energetic mediocrity has managed to supplant modest superiority, misunderstood because of its modesty; yet, and notwithstanding, initiation being an essential law of religious life, a society which is instinctively magical formed at the decline of the pontifical power, and speedily concentrated in itself alone the whole strength of Christianity, because, though it only understood vaguely, it exercised positively the hierarchic power resident in the ordeals of initiation, and the omnipotence of faith in passive obedience.

What, in fact, did the candidate in the old initiations? He entirely abandoned his life and liberty to the masters of the temples of Thebes or Memphis; he advanced resolutely through unnumbered terrors, which might have led him to imagine that there was a premeditated outrage intended against him; he ascended funeral pyres, swam torrents of black and raging water, hung by unknown counterpoises over unfathomed precipices. . . Was not all this a blind obedience in the full force of the term? Is it not the most absolute exercise of liberty to abjure liberty for a time so that we may attain emancipation? Now, this is precisely what must be done, and what has been done invariably, by those who aspire to the *sanctum regnum* of magical omnipotence. The disciples of Pythagoras condemned themselves to inexorable silence for many years; even the sectaries of Epicurus only comprehended the sovereignty of pleasure by the acquisition of sobriety and calculated temperance. Life is a warfare in which we must give proofs if we would advance; power does not surrender of itself; it must be seized.

Initiation by contest and ordeal is therefore indispensable for the attainment of the practical science of magic. We have already indicated after what manner the four elementary forms may be overcome, and will not repeat it here; we refer those of our readers who would inquire into the ceremonies of ancient initiations to the works of Baron Tschoudy, author of the "Blazing Star," "Adonhiramite Masonry," and some other most valuable masonic treatises.

Here we must insist upon a reflection, namely, that the intellectual and social chaos in the midst of which we are perishing, has been caused by the neglect of initiation, with its ordeals and its mysteries. Men, whose zeal was greater than their science, carried away by the popular maxims of the Gospel, came to believe in the primitive and absolute equality of men. A famous *halluciné*, the eloquent and unfortunate Rousseau, propagated this paradox with all the magic of his style – that society alone depraves men – much as if he had said that competition and emulation in labor render workmen idle. The essential law of nature, that of initiation by works and of voluntary and toilsome progress, has been fatally misconstrued; masonry has had its deserters, as Catholicism its apostates. What has been the consequence? The substitution of the steel plane for the intellectual and symbolical plane. To preach equality to what is beneath, without instructing it how to rise upward, is not this binding us to descend ourselves? And hence we have descended to the reign of the carmagnola, the sanscullotes, and Marat. To restore tottering and distracted society, the hierarchy and initiation must be again established. The task is difficult, but the whole intelligent world feels that it is necessary to undertake it. Must we pass through another deluge before succeeding? We trust earnestly not, and this book, perhaps the greatest but not the last of our audacities, is an appeal unto all that is yet alive for the reconstitution of life in the very middle of decomposition and death.

Chapter X: The Key of Occultism

Let us now examine the question of pantacles, for all magical virtue is there, since the secret of force is in the intelligence which directs. We have already given the symbol and interpretation of the pantacles of Pythagoras and Ezekiel, so that we have no need to recur to these; we shall prove in a later chapter that all the instruments of Hebrew worship were pantacles, and that the first and final word of the Bible was written in gold and in brass by Moses, in the tabernacle and on all its accessories. But each magus can and should have his individual pantacle, for, understood accurately, a pantacle is the perfect summary of a mind. Hence, we find in the magical calendars of Tycho Brahe and Duchentau, the pantacles of Adam, Job, Jeremiah, Isaiah, and of all the other great prophets who have been, each in his turn, the kings of the Kabbalah and the grand rabbins of science.

The pantacle, being a complete and perfect synthesis, expressed by a single sign, serves to focus all intellectual force into a glance, a recollection, a touch. It is, so to speak, a starting-point for the efficient projection of the will. Necromancers and goëtic magicians traced their infernal pantacles on the skin of the victims they immolated. The sacrificial ceremonies, the manner of skinning the kid, then of salting, drying and whitening the skin, are given in a number of clavicles and grimoires. Some Hebrew kabbalists fell into similar follies, forgetting the anathemas pronounced in the Bible against those who sacrifice on high places or in the caverns of the earth. All spilling of blood operated ceremonially is abominable and impious, and since the death of Adonhiram the Society of true Adepts has a horror of blood – *Ecclesia abhorret à sanguine.*

The initiatory symbolism of pantacles adopted throughout the East is the key of all ancient and modern mythologies. Apart from knowledge of the hieroglyphic alphabet, one would be lost among the obscurities of the Vedas, the Zend-Avesta and the Bible. The tree which brings forth good and evil, the source of the four rivers, one of which waters the land of gold, that is, of light, and another flows through Ethiopia, or the kingdom of darkness; the magnetic serpent who seduces the woman, and the woman who seduces the man, thus making known the law of attraction; subsequently the Cherub or Sphinx placed at the gate of the Edenic sanctuary, with the fiery sword of the guardians of the symbol; then regeneration by labor and propagation by sorrow, which is the law of initiations and ordeals; the division of Cain and Abel, which is the same symbol as the strife of Anteros and Eros; the ark borne upon the waters of the deluge like the coffer of Osiris; the black raven who does not return and the white dove who does, a new setting forth of the dogma of antagonism and balance – all these magnificent kabbalistic allegories of Genesis, which, taken literally and accepted as actual histories, merit even more derision and contempt than Voltaire heaped upon them, become luminous for the initiate, who still hails with enthusiasm and love the perpetuity of the true doctrine and the universality of initiation identical in all sanctuaries of the world.

The five books of Moses, the Prophecy of Ezekiel, and the Apocalypse of St. John are the three kabbalistic keys of the whole Biblical edifice. The sphinxes of Ezekiel are identical with those of the sanctuary and the ark, and are a quadruple reproduction of the Egyptian tetrad; the wheels revolving in one another are the harmonious spheres of Pythagoras; the new temple, the plan of which is given according to wholly kabbalistic measures, is the type of the labors of primitive masonry. St. John, in his Apocalypse, reproduces the same images and the same numbers, and reconstructs the Edenic world ideally in the New Jerusalem; but at the source of the four rivers the solar lamb replaces the mysterious tree. Initiation by toil and blood has been accomplished, and there is no more temple because the light of truth is universally diffused, and the world has become the temple of justice. This splendid final vision of the Holy Scriptures, this divine Utopia which the Church has referred with good reason for its realization to a better life, has been the pitfall of all ancient arch-heretics and of many modern idealists. The simultaneous emancipation and absolute equality of all men involve the arrest of progress and consequently of life; in a world where all are equal there could no longer be infants or the aged; birth and death could not therefore be admitted. This is sufficient to demonstrate that the New Jerusalem is no more of this world than the primeval paradise, wherein there was no knowledge of good or evil, of liberty, of generation, or of death. The cycle of our religious symbolism begins and ends therefore in eternity.

Dupuis and Volney lavished their great erudition to discover this relative identity of all symbols, and arrived at the negation of every religion. We attain by the same path to a diametrically opposed affirmation, and we recognize with admiration that there have never been any false religions in the civilized world; that the divine light, the splendor of the supreme reason of the Logos, of that word which enlightens every man coming into the world, has been no more wanting to the children of Zoroaster than to the faithful sheep of St. Peter; that the permanent, the one, the universal revelation, is written in visible nature, explained in reason, and completed by the wise analogies of faith; that there is, finally, but one true religion, one doctrine, and one legitimate belief, even as there is but one God, one reason, and one universe; that revelation is obscure for no one, since the whole world understands more or less both truth and justice, and since all that is possible can only exist analogically to all that is. BEING IS BEING, אהיה אשר אהיה.

The apparently bizarre figures presented by the Apocalypse of St. John are hieroglyphics, like those of all oriental mythologies, and can be comprised in a series of pantacles. The initiator, clothed in white, standing between seven golden candlesticks and holding seven stars in his hand, represents the unique doctrine of Hermes and the universal analogies of the light. The woman clothed with the sun and crowned with twelve stars is the celestial Isis, or the gnosis; the serpent of material life seeks to devour her child, but she takes unto herself the wings of the eagle and flies away into the desert – a protestation of the prophetic spirit against the materialism of official religion. The mighty angel with the face of a sun, a rainbow for nimbus, and a cloud for vestment, having pillars of fire for his legs, and setting one foot upon the earth and another on the sea, is truly a kabbalistic Panthea. His feet represent the equilibrium of BRIAH, or the world of forms; his legs are the two pillars of the Masonic temple, JAKIN and BOAZ; his body, veiled by clouds, from which issues a hand holding a book, is the sphere of JETZIRAH, or initiatory ordeals; his solar head, crowned with the radiant septenary, is the world of ATZILUTH, or perfect revelation;

and we can only be excessively astonished that Hebrew kabbalists have not recognized and made known this symbolism, which so closely and inseparably connects the highest mysteries of Christianity with the secret but invariable doctrine of all the masters in Israel. The beast with seven heads, in the symbolism of St. John, is the material and antagonistic negation of the luminous septenary; the Babylonian harlot corresponds after the same manner to the woman clothed with the sun; the four horsemen are analogous to the four allegorical animals; the seven angels with their seven trumpets, seven cups, and seven swords characterize the absolute of the struggle of good against evil by speech, by religious association, and by force. Thus, are the seven seals of the occult book successively opened, and universal initiation is accomplished. The commentators who have sought anything else in this book of the transcendent Kabbalah have lost their time and their trouble only to make themselves ridiculous. To discover Napoleon in the angel Apollyon, Luther in the star which falls from heaven, Voltaire or Rousseau in the grasshoppers armed like warriors, is merely high fantasy. It is the same with all the violence done to the names of celebrated persons so as to make them numerically equivalent to that fatal number 666, which we have already sufficiently explained. When we think that men like Bossuet and Newton amused themselves with such chimeras, we can understand that humanity is not so malicious in its nature as might be supposed from the complexion of its vices.

Chapter XI: The Triple Chain

The great work in practical magic, after the education of the will and the personal creation of the magus, is the formation of the magnetic chain, and this secret is truly that of priesthood and of royalty. To form the magnetic chain is to originate a current of ideas which produces faith and draws a large number of wills in a given circle of active manifestation. A well-formed chain is like a whirlpool which sucks down and absorbs all. The chain may be established in three ways – by signs, by speech and by contact. The first is by inducing opinion to adopt some sign as the representation of a force. Thus, all Christians communicate by the sign of the cross, masons by that of the square beneath the sun, the magi by that of the microcosm, made by extending the five fingers, etc. Once accepted and propagated, signs acquire force of themselves. In the early centuries of our era, the sight and making of the sign of the cross was enough to make proselytes to Christianity. What is called the miraculous medal continues in our own days to affect a great number of conversions by the same magnetic law. The vision and illumination of the young Israelite, Alphonse de Ratisbonne, is the most remarkable fact of this kind. Imagination is creative not only within us but without us by means of our fluidic projections, and undoubtedly the phenomena of the

Transcendent Magic

labarum of Constantine and the cross of Migné should be attributed to no other cause.

The magic chain of speech was typified among the ancients by chains of gold, which issued from the mouth of Hermes. Nothing equals the electricity of eloquence. Speech creates the highest intelligence in the most grossly constituted masses. Even those who are too remote for actual hearing understand by excitement, and are carried away with the crowd. Peter the Hermit convulsed Europe by his cry of "God wills it!" A single word of the Emperor electrified his army and made France invincible. Proudhon destroyed socialism by his celebrated paradox: "Property is robbery." A current saying is frequently sufficient to overturn a reigning power. Voltaire knew this well – who shook the world by sarcasms. So, also, he who feared neither pope nor king, neither parliament nor Bastille, was afraid of a pun. We are on the verge of accomplishing the intentions of that man whose sayings we repeat.

The third method of establishing the magic chain is by contact. Between persons who meet frequently, the head of the current soon manifests, and the strongest will is not slow to absorb the others. The direct and positive grasp of hand by hand completes the harmony of dispositions, and it is for this reason a mark of sympathy and intimacy. Children, who are guided instinctively by nature, form the magic chain by playing at bars or rounds; then gaiety spreads, then laughter rings. Circular tables are more favorable to convivial feasts than those of any other shape. The great circular dance of the Sabbath, which concluded the mysterious assemblies of adepts in the middle ages, was a magic chain, which joined all in the same intentions and the same acts. It was formed by standing back to back and linking hands, the face outside the circle, in imitation of those antique sacred dances, representations of which are still found on the sculptures of old temples. The electric furs of the lynx, panther and even domestic cat, were stitched to their garments, in imitation of the ancient bacchanalia; hence comes the tradition that the Sabbath miscreants each wore a cat hung from the girdle, and that they danced in this guise.

The phenomena of tilting and talking tables has been a fortuitous exhibition of fluidic communication by means of the circular chain. Mystification combined with it afterwards, and even educated and intelligent persons were so infatuated with the novelty that they hoaxed themselves, and became the dupes of their own absurdity. The oracles of the tables were answers more or less voluntarily suggested or extracted by chance; they resembled the conversations which we hold or hear in dreams. Other and stranger phenomena may have been the external manifestations of imaginations operating in common. We, however, by no means deny the possible intervention of elementary spirits in these occurrences, as in those of divination by cards or by dreams; but we do not believe that it has been in any sense proven, and we are therefore in no way obliged to admit it.

One of the most extraordinary powers of human imagination is the fulfilment of the desires of the will, or even of its apprehensions and fears. We believe easily anything that we fear or desire, says a proverb; and it is true, because desire and fear impart to imagination a realizing power, the effects of which are incalculable. How is one attacked, for example, by a disease about which one feels nervous? We have already cited the opinions of Paracelsus on this point, and have established in our doctrinal part the occult laws confirmed by experience; but in magnetic currents, and by mediation of the chain, the realizations are all the more strange because almost invariably unexpected, at least when the chain has not been formed by an intelligent, sympathetic, and powerful leader. In fact, they are the result of purely blind and fortuitous combinations. The vulgar fear of superstitious feasters, when they find themselves thirteen at table, and their conviction that some misfortune threatens the youngest and weakest among them, is, like most superstitions, a remnant of magical science. The duodenary being a complete and cyclic number in the universal analogies of nature, invariably attracts and absorbs the thirteenth, which is regarded as a sinister and superfluous number. If the grindstone of a mill be represented by the number twelve, then thirteen is that of the grain which is to be ground. On kindred considerations, the ancients established the distinctions between lucky and unlucky numbers, whence came the observance of days of good or evil augury. It is in such concerns, above all, that imagination is creative, so that both days and numbers seldom fail to be propitious or otherwise to those who believe in their influence. Consequently, Christianity was right in proscribing the divinatory sciences, for in thus diminishing the number of blind chances, it gave further scope and empire to liberty.

Printing is an admirable instrument for the formation of the magic chain by the extension of speech. No book is lost; as a fact, writings go invariably precisely where they should go, and the aspirations of thought attract speech. We have proved this a hundred times in the course of our magical initiation; the rarest books have offered themselves without seeking as soon as they became indispensable. Thus, have we recovered intact that universal science which so many learned persons have regarded as engulfed by a number of successive cataclysms; thus, have we entered the great magical chain which began with Hermes or Enoch, and will only end with the world. Thus, have we been able to evoke, and come face to face with, the spirits of Apollonius, Plotinus, Synesius, Paracelsus, Cardanus, Cornelius Agrippa and others less or more known, but too religiously celebrated to make it possible for them to be named lightly. We continue their great work, which others will take up after us. But unto whom shall it be given to complete it?

Chapter XII: The Great Work

To be ever rich, to be always young and to die never: such, from all time, has been the dream of alchemists. To change lead, mercury, and all other metals into gold, to possess the universal medicine and the elixir of life – such is the problem which must be solved to accomplish this desire and to realize this dream. Like all magical mysteries, the secrets of the great work have a triple meaning; they are religious, philosophical and natural. The philosophical gold in religion is the absolute and supreme reason; in philosophy, it is truth; in visible nature, it is the sun: in the subterranean and mineral world, it is the purest and most perfect gold. Hence the search after the great work is called the search for the absolute, and this work itself is termed the operation of the sun. All masters of science recognize that it is impossible to achieve material results until we have found all the analogies of the universal medicine and the philosophical stone in the two superior degrees. Then, it is affirmed, is the labor simple, light and inexpensive: otherwise, it consumes to no purpose the life and fortune of the bellows-blower.

The universal medicine is, for the soul, supreme reason and absolute justice; for the mind, it is mathematical and practical truth; for the body, it is the quintessence, which is a combination of gold and light. In the superior world, the first matter of the great work is enthusiasm and activity; in the intermediate world, it is intelligence and industry; in the inferior world, it is labor; in science it is sulphur, mercury and salt, which, volatilized and fixed alternately, compose the Azoth of the sages. Sulphur corresponds to the elementary form of fire, mercury to air and water, salt to earth. All the masters in alchemy who have written concerning the great work have employed symbolical and figurative expressions, and have rightly done so, as much to deter the profane from operations which would, for them, be dangerous, as to make themselves intelligible to adepts, by revealing the entire world of analogies which is ruled by the one and sovereign dogma of Hermes. For such, gold and silver are the sun and moon, or the king and queen; sulphur is the flying eagle; mercury is the winged and bearded hermaphrodite, throned upon a cube and crowned with flames; matter or salt is the winged dragon; metals in the molten state are lions of various colors; finally, the whole work is symbolized by the pelican and phœnix. Hermetic art is, therefore, at one and the same time, a religion, a philosophy and a natural science. Considered as religion, it is that of the ancient magi and the initiates of all the ages; as a philosophy, its principles may be found in the school of Alexandria and in the theories of Pythagoras; as science, its principles must be sought from Paracelsus, Nicholas Flamel, and Raymund Lully. The science is true only for those who accept and understand the philosophy and religion, and its processes are successful only for the adept who has attained sovereign volition, and has thus become the monarch of the elementary world, for the great agent of the solar work is that force described in the Hermetic symbol of the Emerald Table: it is universal magical power; it is the igneous spiritual motor; it is the

Transcendent Magic

Od of the Hebrews, and the astral light, according to the expression which we have adopted in this work. There is the secret, living and philosophical fire, of which all Hermetic philosophers speak only with the most mysterious reservations; there is the universal sperm, the secret of which they guarded, representing it only under the emblem of the caduceus of Hermes. Here then is the great Hermetic arcanum, and we reveal it for the first time clearly and devoid of mystical figures; what the adepts term dead substances are bodies as found in nature; living substances are those which have been assimilated and *magnetized* by the science and will of the operator. Therefore, the great work is something more than a chemical operation; it is an actual creation of the human Word initiated into the power of the Word of God Himself.

הדאבד :
הנתיב הל א נקרי שבל תמידי
כי הוא המנהיג דהשמש והירה
ושאר הבובבים והצוורות בל
אהד מרום בנלו ונותז לבל
הנבראים ממעורבתם אל
המזלות והצוורות :

This Hebrew text, which we transcribe in proof of the authenticity and reality of our discovery, is derived from the rabbinical Jew Abraham, the master of Nicholas Flamel, and it is found in his occult commentary on the Sepher Jetzirah, the sacred book of the Kabbalah. This commentary is extremely rare, but the sympathetic potencies of our chain led us to the discovery of a copy which has been preserved since the year 1643 in the Protestant church at Rouen. On its first page there is written: *Ex dono*, then an illegible name: *Dei magni*.

The creation of gold in the great work takes place by transmutation and multiplication. Raymund Lully states that in order to make gold we must have gold and mercury, while in order to make silver we must have silver and mercury. Then he adds: "By Mercury, I understand that mineral spirit which is so refined and purified that it gilds the seed of gold and silvers the seed of silver." Doubtless he is here speaking of Od, or Astral Light. Salt and sulphur are serviceable in the work only for the preparation of mercury; it is with mercury above all that the magnetic agent must be assimilated and incorporated. Paracelsus, Raymund Lully and Nicholas Flamel seem alone to have perfectly understood this mystery. Basil Valentine and Trevisan indicate it after an incomplete manner, which might be capable of another interpretation. But the most curious things which we have found on this subject are indicated by the mystical figures and magical legends in a book of Henry Khunrath, entitled *Amphitheatrum Sapientæ Æternæ*. Khunrath represents and resumes the most learned Gnostic schools, and connects in symbology with the mysticism of Synesius. He affects Christianity in expressions and in signs, but it is easy to see that his Christ is the Abraxas, the luminous pentagram radiating on the astronomical cross, the incarnation in humanity of the sovereign sun celebrated by the Emperor Julian; it is the luminous and living manifestation of that Ruach-Elohim which, according to Moses, brooded and worked upon the bosom of the waters at the birth of the world; it is the man-sun, the monarch of light, the supreme magus, the master and conqueror of the serpent, and in the fourfold legend of the evangelists, Khunrath finds the allegorical key of the great work. One of the pantacles of his magical book represents the philosophical stone erected in the middle of a fortress surrounded by a wall in which there are twenty impracticable gates. One alone conducts to the sanctuary of the great work. Above the stone there is a triangle placed upon a winged dragon, and on the Stone is graven the name of Christ qualified as the symbolical image of all Nature. "It is by him alone," he adds, "that thou canst obtain the universal medicine for men, animals, vegetables and minerals." The winged dragon, ruled by the triangle, represents, therefore, the Christ of Khunrath; that is, the sovereign intelligence of light and life; it is the secret of the pentagram; it is the highest dogmatic and practical mystery of traditional magic. Thence unto the grand and ever-incommunicable maxim there is only one step.

The kabbalistic figures of Abraham the Jew, which imparted to Flamel the first desire for knowledge, are no other than the twenty-two keys of the Tarot, imitated and resumed in the twelve Keys of Basil Valentine. There the sun and moon reappear under the figures of emperor and empress; Mercury is the juggler; the Great Hierophant is the adept or abstractor of the quintessence; death, judgement, love, the dragon or devil, the hermit or lame elder and, finally, all the remaining symbols are there found with their chief attributes, and almost in the same order. It could have scarcely been otherwise, since the Tarot is the primeval book and the keystone of the occult sciences; it must be Hermetic, because it is kabbalistic, magical and theosophical. So, also, we find in the combination of its twelfth and twenty-second Keys, superposed one upon the other, the hieroglyphic revelation of the solution of the grand work and its mysteries. The twelfth key represents a man hanging by one foot from a gibbet composed of three trees or posts, forming the Hebrew letter ה; the man's arms and head constitute a triangle with his head, and his entire hieroglyphical shape is that of a reversed triangle surmounted by a cross, an alchemical symbol known to all adepts, and representing the accomplishment of the great work. The twenty-second key, which bears the number twenty-one, because the fool which precedes it carries no numeral, represents a youthful female divinity slightly veiled and running in a flowering circle, supported at four corners by the four beasts of the Kabbalah. In the Italian Tarot this divinity has a rod in either hand; in the Besançon Tarot, the two wands are in one hand while the other is placed upon her thigh, both equally remarkable symbols of magnetic action, either alternate in its polarization, or simultaneous by opposition and transmission.

The great work of Hermes is, therefore, an essentially magical operation, and the highest of all, for it supposes the absolute in science and volition. There is light in gold, gold in light, and light in all things. The intelligent will, which assimilates the light, directs in this manner the operations of substantial form, and uses chemistry solely as a secondary instrument. The influence of human will and intelligence upon the operations of nature, dependent in part on its labor, is otherwise a fact so real that all serious alchemists have succeeded in proportion to their knowledge and their faith, and have reproduced their thought in the phenomena of the fusion, salification, and recomposition of metals. Agrippa, who was a man of immense erudition and fine genius, but pure philosopher and sceptic, could not transcend the limits of metallic analysis and synthesis. Etteilla, a confused, obscure, fantastic but persevering kabbalist, reproduced in alchemy the eccentricities of his misconstrued and mutilated Tarot; metals in his crucibles assumed extraordinary forms, which excited the curiosity of all Paris, with no greater profit to the operator than the fees which were paid by his visitors. An obscure bellows-blower of our own time, who died mad, poor Louis Cambriel, really cured his neighbors, and, by the evidence of all his parish, brought back to life a smith who was his friend. For him the metallic work took the most inconceivable and apparently illogical forms. One day he beheld the figure of God himself in his crucible, incandescent like the sun, transparent as crystal, his body composed of triangular conglomerations, which Cambriel naïvely compared to quantities of tiny pears.

One of our friends, who is a learned kabbalist, but belongs to an initiation which we regard as erroneous, performed recently the chemical operations of the great work and succeeded in weakening his eyes through the excessive brilliance of the Athanor. He created a new metal which resembles gold, but is not gold, and hence has no value. Raymond Lully, Nicholas Flamel, and most probably Henry Khunrath made true gold, nor did they take away their secret with them, for it is enclosed in their symbols, and they have further indicated the sources from which they drew for its discovery and for the realization of its effects. It is this same secret which we now ourselves make public.

Chapter XIII: Necromancy

We have boldly declared our opinion, or rather our conviction, as to the possibility of resurrection in certain cases; it remains for us now to complete the revelation of this arcanum and to expose its practice. Death is a phantom of ignorance; it does not exist; everything in nature is living, and it is because it is alive that everything is in motion and undergoes incessant change of form. Old age is the beginning of regeneration, it is the labor of renewing life; and the ancients represented the mystery we term death by the Fountain of Youth, which was entered in decrepitude and left in new childhood. The body is a garment of the soul. When this garment is completely worn out, or seriously and irreparably rent, it is abandoned and never rejoined. But when it is removed by some accident without being worn out or destroyed, it can, in certain cases, be put on again, either by our own efforts or by the assistance of a stronger and more active will than ours. Death is neither the end of life nor the beginning of immortality; it is the continuation and transformation of life. Now a transformation being always a progress, few of those who are apparently dead will consent to return to life, that is, to take up the vestment which they have left behind. It is this which makes resurrection one of the hardest works of the highest initiation, and hence its success is never infallible, but must be regarded almost invariably as accidental and unexpected.

To raise up a dead person we must suddenly and energetically rebind the most powerful chains of attraction which connect it with the body that it has just quitted. It is, therefore, necessary to be acquainted previously with this chain, then to seize thereon, finally to produce an effort of will sufficiently powerful to instantaneously and irresistibly relink it. All this, as we say, is extremely difficult, but is in no sense absolutely impossible. The prejudices of materialistic science exclude resurrection at present from the natural order, and hence there is a disposition to explain all phenomena of this class by lethargies, more or less complicated with signs of death and more or less long in duration. If Lazarus rose again before our doctors, they would record in their memorials to official academies a strange case of lethargy accompanied by an apparent beginning of putrefaction and a strong corpse-like odor; the exceptional occurrence would be labelled with a becoming name, and the matter would be at an end. We have no wish to frighten anyone, and if, out of respect for men with diplomas who represent science officially, it is requisite to term our theories concerning resurrection the art of curing exceptional and aggravated trances, nothing, I hope, will hinder us from making such a concession. But if ever a resurrection has taken place in the world, it is incontestable that resurrection is possible. Now, constituted bodies protect religion, and religion asserts positively the fact of resurrections; therefore, resurrections are possible. From this escape is difficult. To say that they are possible outside the laws of nature, and by an influence contrary to universal harmony, is to affirm that the spirit of disorder, darkness and death, can be the sovereign arbiter of life. Let us not dispute with the worshippers of the devil, but pass on.

It is not religion alone which attests the facts of resurrection; we have collected a number of cases. An occurrence which impressed the imagination of Greuze, the painter, has been reproduced by him in one of his most remarkable pictures. An unworthy son, present at his father's deathbed, seizes and destroys a will unfavorable to himself; the father rallies, leaps up, curses his son and then drops back dead a second time. An analogous and more recent fact has been certified to ourselves by ocular witnesses: a friend, betraying the confidence of one who had just died, tore up a trust-deed he had signed, whereupon the dead person rose up, and lived to defend the rights of his chosen heirs, which this false friend sought to set aside; the guilty person went mad, and the risen man compassionately allowed him a pension. When the Savior raised up the daughter of Jairus, He was alone with three faithful and favored disciples; He dismissed the noisy and loud mourners, saying: "The girl is not dead but sleeping." Then, in the presence only of the father, the mother and the three disciples, that is to say, in a perfect circle of confidence and desire, He took the child's hand, drew her abruptly up and cried to her: "Young girl, I say to thee, arise!" The undecided soul, doubtless in the immediate vicinity of the body, and possibly regretting its extreme youth and beauty, was surprised by the accents of that voice, which was heard by her father and mother trembling with hope, and on their knees; it returned into the body; the maiden opened her eyes, rose up and the Master commanded immediately that food should be given her, so that the functions of life might begin a new cycle of absorption and regeneration. The history of Eliseus, raising up the daughter of the Shunamite, and of St. Paul raising Eutychus, are facts of the same order; the resurrection of Dorcas by St. Peter, narrated so simply in the Acts of the Apostles, is also a history the truth of which can scarcely be reasonably questioned. Apollonius of Tyana seems also to have accomplished similar miracles, and we ourselves have been the witness of facts which are not wanting in analogy with these, but the spirit of the century in which we live imposes in this respect the most careful reserve upon us, the thaumaturge being liable to a very indifferent reception at the hands of a discerning public – all which does not hinder the earth from revolving, or Galileo from having been a great man.

The resurrection of a dead person is the masterpiece of magnetism, because it needs for its accomplishment the exercise of a kind of sympathetic omnipotence. It is possible in the case of death by congestion, by suffocation, by exhaustion, or by hysteria. Eutychus, who was resuscitated by St. Paul after falling from a third story, was doubtless not seriously injured internally, but had succumbed to asphyxia, occasioned by the rush of air during his fall, or alternatively to violent shock and to terror. In a parallel case, he who feels conscious of the power and faith necessary for such an accomplishment must, like the apostle, practice insufflation, mouth to mouth, combined with contact of the extremities for the restoration of warmth. Were it simply a matter of what the ignorant call miracle, Elias and St. Paul, who made use of the same procedure, would have spoken in the name of Jehovah or of Christ. It is occasionally enough to take the person by the hand, and raise them quickly, calling them in a loud voice. This procedure, which commonly succeeds in swoons, may even have effect upon the dead, when the magnetizer who exercises it is endowed with speech powerfully sympathetic and possesses what may be called eloquence of tone. He must also be tenderly loved or greatly respected by the person on whom he would operate, and he must perform the work with a great burst of faith and will, which we do not always find ourselves to possess in the first shock of a great sorrow.

What is vulgarly called necromancy has nothing in common with resurrection, and it is at least highly doubtful that in operations connected with this application of magical power, we really come into correspondence with the souls of the dead whom we evoke. There are two kinds of necromancy, that of light and that of darkness, the evocation by prayer, pantacle, and perfumes, and the evocation by blood, imprecations, and sacrilege. We have only practiced the first, and advise no one to devote themselves to the second. It is certain that the images of the dead do appear to the magnetized persons who evoke them; it is certain also that they never reveal any mysteries of the life beyond. They are beheld as they still exist in the memories of those who knew them, and, doubtless as their reflections have left them impressed on the astral light. When evoked specters reply to questions addressed them, it is always by signs or by interior and imaginary impressions, never with a voice which really strikes the ears, and this is comprehensible enough, for how should a shadow speak? With what instrument could it cause the air to vibrate by impressing it in such a manner as to make distinct sounds? At the same time, electrical contacts are experienced from apparitions, and sometimes appear to be produced by the hand of the phantom, but the phenomena is wholly subjective, and is occasioned solely by the power of imagination and the local wealth of the occult force which we term the astral light. The proof of this is that spirits, or at least the specters pretended to be such, may indeed occasionally touch us, but we cannot touch them, and this is one of the most affrighting characteristics of these apparitions, which are at times so real in appearance that we cannot unmoved feel the hand pass through that which seems a body without touching or meeting anything.

We read in ecclesiastical historians that Spiridion, Bishop of

Tremithonte, afterwards invoked as a saint, called up the spirit of his daughter, Irene, to ascertain from her the whereabouts of some concealed money which she had taken in charge for a traveler. Swedenborg communicated habitually with the so-called dead, whose forms appeared to him in the astral light. Several credible persons of our acquaintance have assured us that they have been revisited for years by the dead who were dear to them. The celebrated atheist Sylvanus Maréchal appeared to his widow and one of her friends, to acquaint her concerning a sum of 1,500 francs which he had concealed in a secret drawer. This anecdote was related to us by an old friend of the family.

Evocations should have always a motive and a becoming end; otherwise, they are works of darkness and folly, most dangerous for health and reason. To evoke out of pure curiosity, and to find out whether we shall see anything, is to court fruitless fatigue. The transcendental sciences admit of neither doubt nor puerility. The permissible motive of an evocation may be either love or intelligence. Evocations of love require less apparatus and are in every respect easier. The procedure is as follows: We must, in the first place, carefully collect the memorials of him (or her) whom we desire to behold, the articles he used, and on which his impression remains; we must also prepare an apartment in in which the person lived, or otherwise one of similar kind, and place his portrait veiled in white therein, surrounded with his favorite flowers, which must be renewed daily. A fixed date must then be observed, either the birthday of the person, or one that was most fortunate for his and our own affection, one of which we may believe that his soul, however blessed elsewhere, cannot lose the remembrance; this must be the day of evocation, and we must provide for it during the space of fourteen days. Throughout this period we must refrain from extending to any one the same proofs of affection which we have the right to expect from the dead; we must observe strict chastity, live in retreat, and take only one modest and light collation daily. Every evening at the same hour we must shut ourselves in the chamber consecrated to the memory of the lamented person, using only one small light, such as that of a funeral lamp or taper. This light should be placed behind us, the portrait should be uncovered, and we should remain before it for an hour, in silence; finally, we should fumigate the apartment with a little good incense, and go out backwards. On the morning of the day fixed for the evocation, we should adorn ourselves as if for a festival, not salute any one first, make but a single repast of bread, wine, and roots, or fruits; the cloth should be white, two covers should be laid, and one portion of the bread broken should be set aside; a little wine should be placed also in the glass of the person whom we design to invoke. The meal must be eaten alone in the chamber of evocations, and in presence of the veiled portrait; it must be all cleared away at the end, except the glass belonging to the dead person, and his portion of bread, which must be placed before the portrait. In the evening, at the hour for the regular visit, we must repair in silence to the chamber, light a clear fire of cypress-wood, and cast incense seven times thereon, pronouncing the name of the person whom we desire to behold. The lamp must then be extinguished, and the fire permitted to die out. On this day the portrait must not be unveiled. When the flame is extinct, put more incense on the ashes and invoke God according to the forms of that religion to which the dead person belonged, and according to the ideas which he himself possessed of God. While making this prayer, we must identify ourselves with the evoked person, speak as he spoke, believe in a sense as he believed; then, after a silence of fifteen minutes, we must speak to him as if he were present, with affection and with faith, praying him to manifest to us. Renew this prayer mentally, covering the face with both hands; then call him thrice with a loud voice; tarry on our knees, the eyes closed or covered, for some minutes; then again call thrice upon him in a sweet and affectionate tone, and slowly open the eyes. Should nothing result, the same experiment must be renewed in the following year, and if necessary a third time, when it is certain that the desired apparition will be obtained, and the longer it has been delayed the more realistic and striking it will be.

Evocations of knowledge and intelligence are made with more solemn ceremonies. If concerned with a celebrated personage, we must meditate for twenty-one days upon his life and writings, form an idea of his appearance, converse with him mentally, and imagine his answers; carry his portrait, or at least his name, about us; follow a vegetable diet for twenty-one days, and a severe fast during the last seven. We must next construct the magical oratory, described in the thirteenth chapter of our *Doctrine*. This oratory must be invariably darkened; but if we operate in the daytime, we may leave a narrow aperture on the side where the sun will shine at the hour of evocation, and place a triangular prism before this opening, and a crystal globe, filled with water, before the prism. If the operation be arranged for night, the magic lamp must be so placed that its single ray shall fall upon the altar smoke. The purpose of these preparations is to furnish the magic agent with elements of corporeal appearance, and to ease as much as possible the tension of imagination, which could not be exalted without danger into the absolute illusion of dream. For the rest, it will be understood easily that a beam of sunlight, or the ray of a lamp, colored variously, and falling upon curling and irregular smoke, can in no way create a perfect image. The chafing-dish containing the sacred fire should be in the center of the oratory, and the altar of perfumes hard by. The operator must turn towards the east to pray, and the west to invoke; he must be either alone or assisted by two persons preserving the strictest silence; he must wear the magical vestments, which we have described in the seventh chapter, and must be crowned with vervain and gold. He should bathe before the operation, and all his under garments must be of the most intact and scrupulous cleanliness. The ceremony should begin with a prayer suited to the genius of the spirit about to be invoked and one which would be approved by himself if he still lived. For example, it would be impossible to evoke Voltaire by reciting prayers in the style of St. Bridget. For the great men of antiquity, we may use the Hymns of Cleanthes or Orpheus, with the adjuration terminating the Golden Verses of Pythagoras. In our evocation of Apollonius, we used the magical philosophy of Patricius for the ritual, containing the doctrines of Zoroaster and the writings of Hermes Trismegistus. We recited the Nuctemeron of Apollonius[24] in Greek with a loud voice, and added the following conjuration:

Vouchsafe to be present, O Father of All, and thou Thrice Mighty Hermes, Conductor of the Dead. Asclepius, son of Hephaistus, Patron of the Healing Art: and thou Osiris, Lord of strength and vigor, do thou thyself be present too. Arnebascenis, Patron of Philosophy, and yet again Asclepius, son of Imuthe, who presidest over poetry . . .

. . . .

Apollonius, Apollonius, Apollonius! Thou teachest the Magic of Zoroaster, son of Oromasdes; and this is the worship of the Gods.

---

[24] The Nuctemeron is included at the end of this book. -pnw

For the evocation of spirits belonging to religions issued from Judaism, the following kabbalistic invocation of Solomon should be used, either in Hebrew, or in any other tongue with which the spirit in question is known to have been familiar:

Powers of the Kingdom, be ye under my left foot and in my right hand! Glory and Eternity, take me by the two shoulders, and direct me in the paths of victory! Mercy and Justice, be ye the equilibrium and splendor of my life! Intelligence and Wisdom, crown me! Spirits of MALCHUTH, lead me betwixt the two pillars upon which rests the whole edifice of the temple! Angels of NETSAH and HOD, establish me upon the cubic stone of JESOD! O GEDULAEL! O GEBURAEL! O TIPHERETH! BINAEL, be ye my love! RUACH HOCHMAEL, be thou my light! Be that which thou are and thou shalt be, O KETHERIEL! *Ishim*, assist me in the name of SADDAÏ! *Cherubim*, be my strength in the name of ADONAÏ! *Beni-Elohim*, be my brethren in the name of the Son, and by the powers of ZEBAOTH! *Eloïm*, do battle for me in the name of TETRAGRAMMATON! *Melachim*, protect me in the name of JOD HE VAU HE! *Seraphim*, cleanse my love in the name of ELVOH! *Hashmalim*, enlighten me with the splendors of ELOÏ and Shechinah! *Aralim*, act! *Ophanim*, revolve and shine! *Haioth ha Kadosh*, cry, speak, roar, bellow! Kadosh, Kadosh, Kadosh, SHADDAÏ, ADONAI, JOTCHABAH, EIEAZEREIE! Hallelu-jah, Hallelu-jah, Hallelujah. Amen. אמן.

It should be remembered, above all in conjurations, that the names of Satan, Beelzebub, Adramelek and others do not designate spiritual unities but legions of impure spirits. "Our name is legion, for we are many," says the spirit of darkness in the Gospel. Number constitutes law, and progress takes place inversely in hell – that is to say, the most advanced in Satanic development, and consequently the most degraded, are the least intelligent and feeblest. as the domain of anarchy. That is to say, the most advanced in Satanic development and consequently the most degraded and the least intelligent and feeblest. Thus, a fatal law drives demons downward when they wish and believe themselves to be ascending. So also, those who term themselves chiefs are the most impotent and despised of all. As to the horde of perverse spirits, they tremble before an unknown, invisible, incomprehensible, capricious, implacable chief, who never explains his laws, whose arm is ever stretched out to strike those who fail to understand him. They give this phantom the names of Baal, Jupiter, and even others more venerable, which cannot, without profanation, be pronounced in hell. But this phantom is only the shadow and remnant of God, disfigured by willful perversity, and persisting in their imagination like a vengeance of justice and a remorse of truth.

When the evoked spirit of light manifests with dejected or irritated countenance, we must offer him a moral sacrifice, that is, be inwardly disposed to renounce whatever offends him; and before leaving the oratory, we must dismiss him, saying: "May peace be with thee! I have not wished to trouble thee; do thou torment me not. I shall labor to improve myself as to anything that vexes thee. I pray, and will still pray, with thee and for thee. Pray thou also both with and for me, and return to thy great slumber, expecting that day when we shall awake together. Silence and adieu!"

We must not close this chapter without giving some details on black magic for the benefit of the curious. The practices of Thessalian sorcerers and Roman Canidias are described by several ancient authors. In the first place, a pit was dug, at the mouth of which they cut the throat of a black sheep; the psyllæ and larvæ presumed to be present, and swarming round to drink the blood, were driven off with the magic sword; the triple Hecate and the infernal gods were evoked, and the phantom whose apparition was desired was called upon three times. In the middle ages, necromancers violated tombs, composing philtres and unguents with the fat and blood of corpses combined with aconite, belladonna and poisonous fungi; they boiled and skimmed these frightful compounds over fires nourished with human bones and crucifixes stolen from churches; they added dust of dried toads and ash of consecrated hosts; they anointed their temples, hands, and breasts with the infernal unguent, traced diabolical pantacles, evoked the dead beneath gibbets or in deserted graveyards. Their howlings were heard from afar, and belated travelers imagined that legions of phantoms rose out of the earth; the very trees, in their eyes, assumed appalling shapes; fiery orbs gleamed in the thickets; frogs in the marshes seemed to echo mysterious words of the Sabbath with croaking voices. It was the magnetism of hallucination and the contagion of madness.

The end of procedure in black magic was to disturb reason and produce the feverish excitement which emboldens to great crimes. The grimoires, once seized and burnt by authority everywhere, are certainly not harmless books. Sacrilege, murder, theft, are indicated or hinted as means to realization in almost all these works. Thus, in the Great Grimoire and its modern version, the Red Dragon, there is a recipe entitled "Composition of Death, or Philosophical Stone," a broth of aqua fortis, copper, arsenic and verdigris. There are also necromantic processes, comprising the tearing up of earth from graves with the nails, dragging out bones, placing them crosswise on the breast, then assisting at midnight mass on Christmas eve, and flying out of the church at the moment of consecration, crying: "Let the dead rise from their tombs!" – then returning to the graveyard, taking a handful of earth nearest to the coffin, running back to the door of the church, which has been alarmed by the clamor, depositing the two bones crosswise, again shouting: "Let the dead rise from their tombs," and then, if we escape being seized and shut up in a mad-house, retiring at a slow pace, and count four thousand five hundred steps in a straight line, which means following a broad road or scaling walls. Having traversed this space, you must lie down upon the earth, place yourself as if in a coffin, and repeat in lugubrious tones: "Let the dead rise from their tombs!" Finally, call thrice on the person whose apparition you desire. No doubt anyone who is mad enough and wicked enough to abandon himself to such operations is predisposed to all chimeras and all phantoms. Hence the recipe of the Grand Grimoire is most efficacious, but we advise none of our readers to have recourse to it.

Chapter XIV: Transmutations

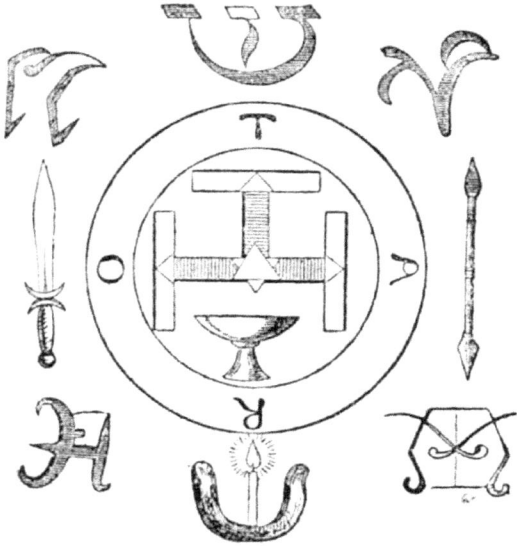

St. Augustus speculates, as we have said, whether Apuleius could have been changed into an ass and then have resumed his human shape. The same doctor might have concerned himself equally with the adventure of the comrades of Ulysses, transformed into swine by Circe. In vulgar opinion, transmutations and metamorphoses have always been the very essence of Magic. Now, the crowd, being the echo of opinion, which is queen of the world, is never perfectly right or entirely wrong. Magic really changes the nature of things, or rather, modifies their appearances at pleasure, according to the strength of the operator's will and the fascination of ambitious adepts. Speech creates its form, and when a person, held infallible, confers a name upon a given thing, he really transforms that thing into the substance signified by the name. The masterpiece of speech and of faith, in this order, is the real transmutation of a substance without change in its appearance. Had Apollonius offered a cup of wine to his disciples, and said to them: "This is my blood, of which ye shall drink hence-forth to perpetuate my life within you;" and had his disciples through centuries believed that they effected transformation by repeating the same words; had they taken the wine, despite its odor and taste, for the real, human, and living blood of Apollonius, we should have to acknowledge this master in theurgy as the most accomplished of enchanters and most potent of all the magi. It would remain for us then to adore him.

Now, it is well known that mesmerists impart for their somnambulists any taste that they choose to plain water, and if we assume a magus having sufficient command over the astral fluid to magnetize at the same moment a whole assembly of persons, otherwise prepared for magnetism by adequate over-excitement, we shall be in a position to explain readily, not indeed the Gospel miracle of Cana, but works of the same class. Are not the fascinations of love, which result from the universal magic of nature, truly prodigious, and do they not actually transform persons and things? Love is a dream of enchantments that transfigures the world; all becomes music and fragrance, all intoxication and felicity. The beloved being is beautiful, is good, is sublime, is infallible, is radiant, glows with health and happiness. When the dream ends we seem to have fallen from the clouds; we are inspired with disgust for the brazen sorceress who took the place of the lovely Melusine, for the Thersites whom we deemed was Achilles or Nereus. What is there we cannot cause the person who loves us to believe? But also, what reason or justice can we instill into those who no longer love us? Love begins magician and ends sorcerer. After creating the illusions of heaven on earth, it realizes those of hell; its hatred is absurd like its ardor, because it is passional, that is, subject to influences which are fatal for it. For this cause it has been proscribed by sages, who declare it the enemy of reason. Are they to be envied or commiserated for thus condemning, doubtless without understanding, the most alluring of ill-doers? All that can be said is that when they spoke thus, they either had not yet loved or they loved no longer.

Things that are external are for us what our word internal makes them. To believe that we are happy is to be happy; whatsoever we esteem becomes precious in proportion to the estimation itself: this is the sense in which we can say that magic changes the nature of things. The Metamorphoses of Ovid are true, but they are allegorical, like the "Golden Ass" of rare Apuleius. The life of beings is a progressive transformation, and its forms can be determined, renewed, prolonged further, or destroyed sooner. If the doctrine of metempsychosis were true, might one not say that the debauch represented by Circe really and materially changes men into swine; for, on this hypothesis, the retribution of vices would be a relapse into animal forms that correspond to them? Now, metempsychosis, which has frequently been misinterpreted, has a perfectly true side; for animal forms communicate their sympathetic impressions to the astral body of man, which speedily reacts on his aptitudes according to the force of his habits. A man of intelligent and passive mildness assumes the inert physiognomy and ways of a sheep, but in somnambulism it is a sheep that is seen, and not a man with a sheepish countenance, as the ecstatic and learned Swedenborg experienced a thousand times. In the kabbalistic book of Daniel the seer, this mystery is represented by the legend of Nebuchodonsor changed into a beast, which, after the common fate of magical allegories, has been mistaken for an actual history. In this way, we can really transform men into animals and animals into men; we can metamorphose plants and alter their virtue; we can endow minerals with ideal properties; it is all a question of willing. We can equally render ourselves visible or invisible at will, and this enables us to explain the mysteries of the ring of Gyges.

In the first place, let us remove from the mind of our readers all supposition of the absurd; that is, of an effect devoid of cause or contradicting its cause. To become invisible one of three things is necessary – the interposition of some opaque medium between the light and our body, or between our body and the eyes of the spectators, or the fascination of the eyes of the spectators in such a manner that they cannot make use of their sight. Of these methods, the third only is magical. Have we not all of us observed that under the government of a strong preoccupation we look without seeing and hurt ourselves against objects in front of us? "So do, that seeing they may not see," said the great Initiator, and the history of this grand master tells us that one day, finding himself on the point of being stoned in the temple, he became invisible and went out. There is no need to repeat here the mystifications of popular grimoires about the ring of invisibility. Some ordain that it shall be composed of fixed mercury, enriched by a small stone which is indispensable to find in a pewit's nest, and kept in a box of the same metal. The author of the "Little Albert" ordains that this ring should be composed of hairs torn from the head of a raging hyena, which recalls the history of the bell of Rodilard. The only writers who have discoursed seriously of the ring of Gyges are Jamblichus, Porphyry, and Peter of Apono. What they say is evidently allegorical, and the representation which they give, or that which can be made from their description, proves that they are really speaking of nothing but the great magical arcanum. One of the figures depicts the universal movement, harmonic and equilibrated in imperishable being; another, which should be formed from an amalgam of the seven metals, calls for a description in detail. It has a double collet and two precious stones – a topaz, constellated under the sign of the sun, and an emerald under the sign of the moon; it should bear on the inner side the occult characters of the planets, and on the outer their known signs, twice repeated and in kabbalistic opposition to each other; that is, five on the right and five on the left; the signs of the sun and moon resuming the four several intelligences of the seven planets. Now, this configuration is no other than that of a pantacle signifying all the mysteries of magical doctrine, and here is the occult significance of the ring: to exercise omnipotence, of which ocular fascination is one of the most difficult demonstrations to give, we must possess all science and know how to make use of it.

Fascination is fulfilled by magnetism. The magus inwardly forbids a whole assembly to see him, and it does not see him. In this manner he passes through guarded gates, and departs from prison in the face of his petrified gaolers. At such times a strange numbness is experienced, and they recall having seen the magus as if in a dream, but never till after he has gone. The whole secret of invisibility, therefore, wholly consists in a power which is capable of definition –that of distracting or paralyzing attention, so that the light reaches the visual organ without impressing the eye of the soul. To exercise this power we must possess a will accustomed to sudden and energetic actions, great presence of mind, and skill no less great in causing diversions among the crowd. Let a man, for example, who is being pursued by his intending murderers, dart into a side street, return immediately, and advance with perfect calmness towards his pursuers, or let him mix with them and seem to be engaged in the chase, and he will certainly make himself invisible. A priest who was being hunted in '93, with the intention of hanging him from a lamp-post, fled down a certain street, assumed a stooping gait, and leaned against a corner, with an intensely preoccupied expression; the crowd of his enemies swept past; not one saw him, or, rather, it never struck anyone to recognize him; it was so unlikely to be he! The person who desires to be seen always makes himself observed, but he who would remain unnoticed effaces himself and disappears. The true ring of Gyges is the will; it is also the rod of transformations, and by its precise and strong formulation it creates the magical word. The omnipotent terms of enchantments are those which express this creative power of forms. The tetragram, which is the supreme word of magic, signifies: "It is that which it shall be," and if we apply it to any transformation whatsoever with full intelligence, it will renew and modify all things, even in the teeth of evidence and common sense. The *hoc est* of the Christian sacrifice is a translation and application of the tetragram; hence this simple utterance operates the most complete, most invisible, most incredible, and most clearly affirmed of all transformations. A still stronger word than that of *transformation* has been judged necessary by councils to express the marvel, that of *transubstantiation*.

The Hebrew terms אמן, אהיה, אגלא, יהוה have been considered by all kabbalists as the keys of magical transformation. The Latin words, *est, sit, esto, fiat*, have the same force when pronounced with full understanding. M. de Montalembert relates seriously, in his legend of St. Elizabeth of Hungary, how one day this saintly lady, surprised by her noble husband, from whom she sought to conceal her good works, in the act of carrying bread to the poor in her apron, told him that she was carrying roses, and it proved on investigation that she had spoken truly; the loaves had been changed into roses. This story is a most gracious magical apologue, and signifies that the truly wise man cannot lie, that the word of wisdom determines the form of things, or even their substance independently of their forms. Why, for example, should not the noble spouse of St. Elizabeth, a good and firm Christian like herself, and believing implicitly in the real presence of the Savior in true human body upon an altar where he beheld only a wheaten host, why should he not believe in the real presence of roses in his wife's apron under the appearances of bread? She exhibited him loaves undoubtedly, but as she had said that they were roses, and as he believed her incapable of the smallest falsehood, he saw and wished to see roses only. This is the secret of the miracle. Another legend narrates how a saint, whose name has escaped me, finding nothing to eat on a Lenten day or a Friday, commanded the fowl to become a fish, and it became a fish. The parable needs no interpretation, and it recalls a beautiful story of St. Spiridion of Tremithonte, the same who evoked the soul of his daughter Irene. One Good Friday a traveler reached the abode of the holy bishop, and as bishops in those days took Christianity in earnest, and were consequently poor, Spiridion, who fasted religiously, had in his house only some salted bacon, which had been made ready for Easter. The stranger was overcome with fatigue and famished with hunger; Spiridion offered him the meat, and himself shared the meal of charity, thus transforming the very flesh which the Jews regard as of all most impure into a feast of penitence, transcending the material law by the spirit of the law itself, and proving himself a true and intelligent disciple of the man-God, who hath established his elect as the monarchs of nature in the three worlds.

Chapter XV: The Sabbath of the Sorcerers

WE return once more to that terrible number fifteen, symbolized in the Tarot by a monster throned upon an altar, mitred and horned, having a woman's breasts and the generative organs of a man – a chimera, a malformed sphinx, a synthesis of monstrosities; below this figure we read a frank and simple inscription – THE DEVIL. Yes, we confront here that phantom of all terrors, the dragon of all theogonies, the Ariman of the Persians, the Typhon of the Egyptians, the Python of the Greeks, the old serpent of the Hebrews, the fantastic monster, the nightmare, the Croquemitaine, the gargoyle, the great beast of the middle ages, and, worse than all this, the Baphomet of the Templars, the bearded idol of the alchemists, the obscene deity of Mendes, the goat of the Sabbath. The frontispiece to this Ritual reproduces the exact figure of the terrible emperor of night, with all his attributes and all his characters.

Let us state now for the edification of the vulgar, for the satisfaction of M. le Comte de Mirville, for the justification of the demonologist Bodin, for the greater glory of the Church, which persecuted Templars, burnt magicians, excommunicated Freemasons, etc. – let us state boldly and precisely that all inferior initiates of the occult science and profaners of the great arcanum, not only did in the past but do now, and will ever, adore what is signified by this alarming symbol. Yes, in our profound conviction, the Grand Masters of the Order of the Templars worshipped the Baphomet, and caused it to be worshipped by their adepts; yes, there existed in the past, and there may be still in the present, assemblies which are presided over by this figure, seated on a throne and having a flaming torch between the horns; but the adorers of this sign do not consider, as do we, that it is the representation of the devil; on the contrary, for them it is that of the god Pan, the god of our modern schools of philosophy, the god of the Alexandrian theurgic school and of our own mystical Neo-platonists, the god of Lamartine and Victor Cousin, the god of Spinoza and Plato, the god of the primitive Gnostic schools; the Christ also of the dissident priesthood; this last qualification, ascribed to the goat of black magic, will not astonish students of religious antiquities who are acquainted with the phases of symbolism and doctrine in their various transformations, whether in India, Egypt or Judea.

The bull, the dog and the goat are the three symbolical animals of Hermetic magic, resuming all the traditions of Egypt and India. The bull represents the earth or salt of the philosophers; the dog is Hermanubis, the Mercury of the sages, fluid, air and water; the goat represents fire, and is at the same time the symbol of generation. Two goats, one pure and one impure, were consecrated in Judea; the first was sacrificed in expiation for sins; the other, loaded with those sins by imprecation, was set at liberty in the desert – a strange ordinance, but one of deep symbolism, signifying reconciliation by sacrifice and expiation by liberty! Now, all the fathers of the Church, who have concerned themselves with Jewish symbolism, have recognized in the immolated goat the figure of him who assumed, as they say, the very form of sin. Hence the Gnostics were not outside symbolical traditions when they gave Christ the Liberator this same mystical figure. All the

Kabbalah and all magic, as a fact, are divided between the cultus of the immolated and that of the emissary goat. There is, therefore, the magic of the sanctuary and that of the wilderness, the white and the black Church, the priesthood of public assemblies and the sanhedrim of the Sabbath. The goat which is represented in our frontispiece bears upon its forehead the sign of the pentagram with one point in the ascendant, which is sufficient to distinguish it as a symbol of the light; he makes the sign of occultism with both hands, pointing upward to the white moon of Chesed, and downward to the black moon of Geburah. This sign expresses the perfect concord between mercy and justice. One of the arms is feminine and the other masculine, as in the androgyne of Khunrath, those attributes we have combined with those of our goat, since they are one and the same symbol. The torch of intelligence burning between the horns is the magical light of universal equilibrium; it is also the type of the soul exalted above matter, even while connected with matter, as the flame connects with the torch. The hideous head of the animal expresses horror of sin, for which the material agent, alone responsible, must alone and forever bear the penalty, because the soul is impassible in its nature, and can suffer only by materializing. The caduceus, which replaces the generative organ, represents eternal life; the scale-covered belly typifies water; the circle above it is the atmosphere, the feathers still higher up signify the volatile; lastly, humanity is depicted by the two breasts and the androgyne arms of this sphinx of the occult sciences. Behold the shadows of the infernal sanctuary dissipated! Behold the sphinx of mediaeval terrors divined and cast from his throne! *Quomodo cedidisti, Lucifer!*

The dread Baphomet henceforth, like all monstrous idols, enigmas of antique science and its dreams, is only an innocent and even pious hieroglyph. How should man adore the beast, since he exercises a sovereign power over it? Let us affirm, for the honor of humanity, that it has never adored dogs and goats any more than lambs or pigeons. In the hieroglyphic orders, why not a goat as much as a lamb? On the sacred stones of Gnostic Christians of the Basilidean sect, are representations of Christ under the diverse figures of kabbalistic animals – sometimes a bird, at others a lion, and, again, a lion or bull-headed serpent; but in all cases He bears invariably the same attributes of light, even as our goat, which cannot be confounded with fabulous images of Satan, owing to his sign of the pentagram.

Let us assert most strongly, to combat the remnants of Manichæanism which are daily appearing among Christians, that as a superior personality and power Satan does not exist. He is the personification of all errors, perversities, and, consequently, of all weaknesses. If God may be defined as He who necessarily exists, may we not define His antagonist and enemy as he who necessarily does not exist? The absolute affirmation of good implies the absolute negation of evil; so also, in the light, shadow itself is luminous. Thus, erring spirits are good to the extent of their participation in being and in truth. There are no shadows without reflections, no nights without moon, phosphorescence, and stars. If he'll be just, it is good. No one has ever blasphemed God. The insults and mockeries addressed to His disfigured images attain Him not.

We have named Manichæanism, and it is by this monstrous heresy that we shall explain the aberrations of black magic. The misconstrued doctrine of Zoroaster and the magical law of two forces constituting universal equilibrium, have caused some illogical minds to imagine a negative divinity, subordinate but hostile to the active divinity. Thus, the impure duad comes into being. Men were mad enough to halve God; the star of Solomon was separated into triangles, and the Manichæans imagined a trinity of night. This evil God, product of sectarian fancies, inspired all manias and all crimes. Sanguinary sacrifices were offered him; monstrous idolatry replaced the true religion; black magic traduced the transcendent and luminous magic of true adepts, and horrible conventicles of sorcerers, ghouls, and stryges took place in caverns and desert places, for dementia soon changes into frenzy, and from human sacrifices to cannibalism there is only one step. The mysteries the Sabbath have been described variously, but they figure always in grimoires and in magical trials. The revelations made on the subject may be classified under three heads: 1. those referring to a fantastic and imaginary Sabbath; 2. those which betray the secrets of the occult assemblies of veritable adepts; 3. revelations of foolish and criminal gatherings, having for their object the operations of black magic. For a large number of unhappy men and women, given over to such mad and abominable practices, the Sabbath was but a prolonged nightmare, where dreams appeared realities and were induced by means of potions, fumigations and narcotic frictions. Baptista Porta, whom we have signalized already as a mystifier, gives in his "Natural Magic," a pretended recipe for the sorcerers' unguent, by means of which they were transported to the Sabbath. It is a composition of child's fat, aconite boiled with poplar leaves and some other drugs, the whole mixed with soot, which could not contribute to the beauty of the naked sorceresses who repaired to the scene anointed with this pomade. There is another and more serious recipe given the same author, which we transcribe in Latin to preserve its grimoire character. *Recipe: suim, acorum vulgare, pentaphyllon, verspertillionis sanguinem solanum somniferum et oleum*, the whole boiled and incorporated to the consistence of an unguent. We infer that compositions containing opiates, the pith of green hemp, Datura stramonium or Laurel-almond, would enter quite as successfully into such preparations. The fat or blood of night-birds added to these narcotics, with black magical ceremonies, would impress imagination and determine the direction of dreams. To Sabbaths dreamed in this manner we must refer the accounts of a goat issuing from pitchers and going back into them after the ceremony; infernal powders obtained from the ordure of this goat, who is called Master Leonard; banquets where abortions are eaten without salt and boiled with serpents and toads; dances, in which monstrous animals or men and women with impossible shapes take part; unbridled debauches where incubi project cold sperm. Nightmare alone could produce or explain such scenes. The unfortunate cure, Gaufridy, and his abandoned penitent Madeline de la Palud, went mad through kindred delusions, and were burned for persisting in affirming them. We must read the depositions of these diseased beings during their trial to understand the extent of the aberration possible to an afflicted imagination. But the Sabbath was not always a dream; it did exist in reality; even now there are secret nocturnal assemblies for the practice of the rites of the old world, some of which assemblies have a religious and social object, while that of others is concerned with orgies and conjurations. From this two-fold point of view we propose to consider the true Sabbath, of the magic of light in the one case and the magic of darkness in the other.

When Christianity proscribed the public exercise of the ancient worships, the partisans of the latter were compelled to meet in secret for the celebration of their mysteries. Initiates presided over these assemblies, and soon established among the varieties of the worships a kind of orthodoxy, more easily facilitated by the aid of magical truth, because proscription unites wills and forges bonds of brotherhood between men. Thus, the Mysteries of Isis, of Ceres Eleusinia, of Bacchus, combined with those of the good goddess and primeval Druidism. The meetings took place usually between the days of Mercury and Jupiter, or between those of Venus and Saturn; the proceedings included the rites of initiation, exchange of mysterious signs, singing of symbolical hymns, the cementing of union at the banqueting-board, the successive formation of the magical chain at table and in the dance; and, finally, the meeting broke up after renewing pledges in the presence of the chiefs and receiving instructions from them. The candidate for the Sabbath was led, or rather carried, to the assembly, his eyes covered by the magical mantle in which he was completely enveloped, he was led between immense fires, while alarming noises were made about him. When his face was bared, he found himself surrounded by infernal monsters, and in the presence of a colossal and hideous goat which he was commanded to adore. All these ceremonies were tests of his force of character and confidence in his initiators. The final ordeal was most decisive of all because it was at first sight humiliating and ridiculous to the mind of the candidate; he was commanded without circumspection to kiss respectfully the posterior of the goat; if he refused, his head was again covered, and he was transported to a distance from the assembly with such extraordinary rapidity that he believed himself whirled through the air; if he assented, he was taken round the symbolical idol, and there found, not a repulsive and obscene object, but the young and gracious countenance of a priestess of Isis or Maia, who gave him a maternal salute, and he was then admitted to the banquet. As to the orgies which in many such assemblies followed the banquet, we must beware of believing that they were generally permitted these secret agapae; at the same time, it is known that a number of gnostic sects practiced them in their conventicles during early centuries of Christianity. That the flesh had its protestants in those ages of asceticism and compression of the senses was inevitable, and can occasion no surprise, but we must not accuse transcendent magic of irregularities which it has never authorized. Isis is chaste in her widowhood; Diana Panthea is a virgin; Hermanubis, possessing both sexes, can satisfy neither; the Hermetic hermaphrodite is pure; Apollonius of Tyana never yielded to the seductions of pleasure; the Emperor Julian was a man of rigid continence; Plotinus of Alexandria was ascetic in the manner of his life; Paracelsus was such a stranger to foolish love that his sex was suspected; Raymund Lully was initiated in the final secrets of science only after a hopeless passion which made him chaste forever. It is also a magical tradition that pantacles and talismans lose all their virtue when he who wears them enters a house of prostitution or commits an adultery. The Sabbath of orgies must not therefore be considered as that of the veritable adepts.

With regard to the term Sabbath, some have traced it to the name of Sabasius, and other etymologies have been imagined. The most simple, in our opinion, connects it with the Jewish Sabbath, for it is certain that the Jews, most faithful depositaries of the secrets of the Kabbalah, were almost invariably the great masters in magic during the middle ages. The Sabbath was therefore the Sunday of Kabbalists, the day of their religious festivals, or rather the night of their regular assembly. This feast, surrounded with mysteries, had the vulgar timidity for its safeguard and escaped persecution by terror. As to the diabolical Sabbath of necromancers, it was a counterfeit of that of the magi, an assembly of malefactors who exploited idiots and fools. There horrible rites were practiced and abominable potions compounded, there sorcerers and sorceresses laid their plans and instructed one another for the common support of their reputation in prophecy and divination; at that period diviners were generally consulted and followed a lucrative profession while exercising a real power. Such institutions neither had nor could possess any regular rites; everything depended on the caprice of the chiefs and the vertigo of the assembly. What was narrated by some who had been present served as a type for all nightmares of hallucination and from this chaos of impossible realities and demoniac dreams have issued the revolting and foolish histories of the Sabbath which figure in magical processes and in the books of such writers as Sprenger, Delancere, Delrio and Bodin.

The rites of the Gnostic Sabbath were imported into Germany by an association which took the name of Mopses. It replaced the Kabbalistic goat by the Hermetic dog, and the candidate, male or female, for the order initiated women, were brought in with eyes bandaged; the same infernal noise was made in their neighborhood, which surrounded the name of Sabbath with so many inexplicable rumors; they were asked whether they were afraid of the devil, and were abruptly required to choose between kissing the posterior of the grand master and that of a small silk-covered figure of a dog, which was substituted for the old grand idol of the goat of Mendes. The sign of recognition was a ridiculous grimace, which recalls the phantasmagoria of the ancient Sabbath and the masks of the assistants. For the rest, their doctrine is summed up in the cultus of love and license. The association came into existence when the Roman Church was persecuting Freemasonry. The Mopses pretended to recruit only among Catholics, and for the oath at reception they substituted a solemn engagement upon honor to reveal no secrets of the order. It was more effectual than any oath, and left nothing for religion to object.

The name of the Templar Baphomet, which should be spelt kabbalistically backwards, is composed of three abbreviations: TEM. OHP. AB., *Templi omnium hominum pacis abbas*, "the father of the temple of peace of all men." According to some, the Baphomet was a monstrous head; according to others, a demon in the form of a goat. A sculptured coffer was disinterred recently in the ruins of an old commandery of the temple, and antiquaries observed upon it a baphometic figure, corresponding by its attributes to the goat of Mendes and the androgyne of Khunrath. It was a bearded figure with a female body, holding the sun in one hand and the moon in the other, attached to chains. Now, this virile head is a beautiful allegory which attributes to thought alone the initiative and creative principle. Here the head represents spirit and the body matter. The orbs enchained to the human form, and directed by that nature of which intelligence is the head, are also magnificently allegorical. The sign all the same was discovered to be obscene and diabolical by the learned men who examined it. Can we be surprised after this at the spread of mediæval superstition in our own day! One thing only surprises me, that, believing in the devil and his agents, men do not rekindle the faggots. M. Veuillot is logical and demands it; one should honor men who have the courage of their opinions.

Pursuing our curious researches, we come now to the most horrible mysteries of the Grimoire, those which are concerned with evocations of devils and pacts with hell. After attributing a real existence to the absolute negation of goodness, after having enthroned the absurd and created a god of falsehood, it remained for human folly to invoke the impossible idol, and these maniacs have done. We were informed lately that the most reverend Father Ventura, formerly Superior of the Theatines, Bishops' Examiner, etc., after reading our Doctrine, declared that the kabbalah was, in his opinion, an invention of the devil, and that the star of Solomon was another diabolical device to persuade the world that Satan was the same as God. See what is taught seriously by the masters in Israel! The ideal of nothingness and night inventing a sublime philosophy which is the universal basis of faith and the keystone of all temples! The demon placing his signature by the side of God's! My venerable masters in theology, you are greater sorcerers than you or others are aware, and He who said: "The devil is a liar like his father," would have had some observations to make on the decisions of your reverences.

Evokers of the devil must before all things belong to a religion which admits a devil, creator and rival of God. To invoke a power, we must believe in it. Given this firm faith in the religion of the devil, we must proceed as follows to enter into correspondence with this pseudo-Deity:

MAGICAL AXIOM.

In the circle of its action, every word creates that which it affirms.

DIRECT CONSEQUENCE.

He who affirms the devil, creates or makes the devil.

Conditions of Success in Infernal Evocations.

1. Invincible obstinacy;
2. A conscience at once hardened to crime and most subject to remorse and fear;
3. Affected or natural ignorance;
4. Blind faith in all that is incredible;
5. A completely false idea of God.

We must afterwards – (a) Profane the ceremonies of the cultus in which we believe; (b) offer a bloody sacrifice; (c) procure the magic fork, which is a branch of a single beam of hazel or almond, cut at one blow with the new knife used for the sacrifice. It must terminate in a fork, which must be armored with iron or steel made from the blade of the before-mentioned knife. A fast of fifteen days must be observed, taking a single unsalted repast after sundown; this repast should consist of black bread and blood seasoned with unsalted spices or black beans and milky and narcotic herbs. We must get drunk every five days, after sundown, on wine in which five heads of black poppies and five ounces of pounded hemp seed have been steeped for five hours, the infusion being strained through a cloth woven by a prostitute; strictly speaking, the first cloth which comes to hand may be used, should it have been woven by a woman. The evocation should be performed on the night between Monday and Tuesday, or that between Friday and Saturday. A solitary and condemned spot must be chosen, such as a cemetery haunted by evil spirits, an avoided ruin in the country, the vaults of an abandoned convent, a place where some murder has been committed, a druidic altar or an old temple of idols. A black seamless and sleeveless robe must be provided; a leaden cap emblazoned with the signs of the moon, Venus, and Saturn; two candles of human fat set in black wooden candlesticks, carved in the shape of a crescent; two crowns of vervain; a magical sword with a black handle; the magical fork; a copper vase containing the blood of the victim; a censer holding perfumes, namely, incense, camphor, aloes, ambergris and storax, kneaded with the blood of a goat, a mole, and a bat; four nails taken from the coffin of an executed criminal; the head of a black cat which has been nourished on human flesh for five days; a bat drowned in blood; the horns of a goat *cum quo puella concubuerit*; and the skull of a parricide.

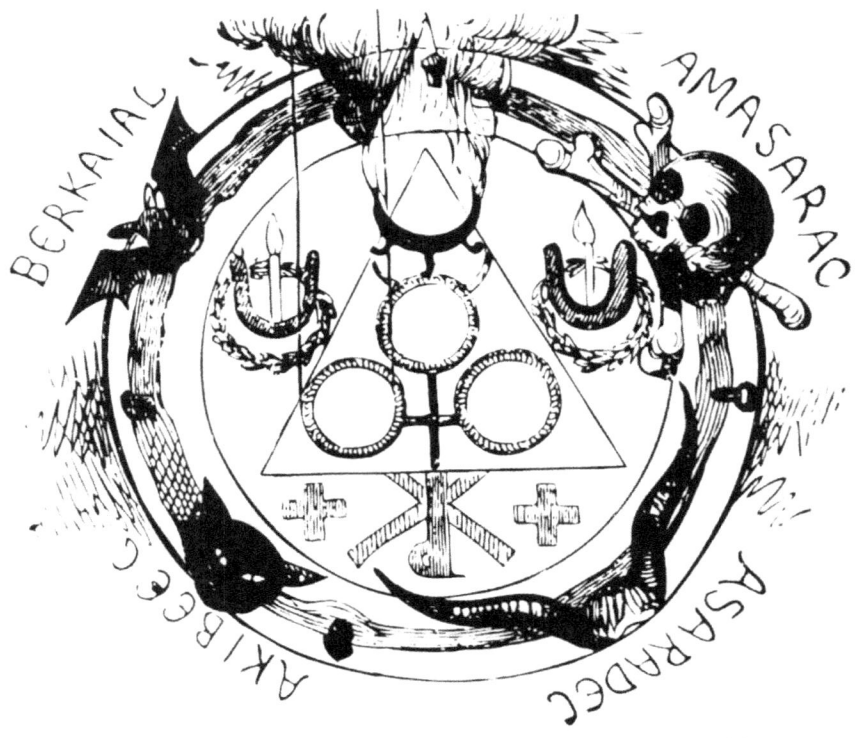

All these hideous and scarcely obtainable objects and having been collected, they must be arranged as follows: A perfect circle is traced by the sword, with a break, or way of issuing, on one side; a triangle is drawn in the circle, and the pantacle thus formed is colored with blood; at one of its angles of the triangle a chafing-dish is placed, and this should have been included among the indispensable objects already enumerated; at the opposite base of the triangle three little circles are described for the operator and his two assistants; behind that of the first the sign of the labarum or monogram of Constantine is drawn, not with the blood of the victim, but with the operator's own blood. The operator and his assistants must have bare feet and covered heads. The skin of the immolated victim must be also brought to the place, and, being cut into strips, must be placed within the circle, forming a second and inner circle, fixed at four corners by the four nails from the coffin already mentioned. Hard by the nails, but outside the circle, must be placed the head of the cat, the human or rather inhuman skull, the horns of the goat, and the bat; they must be sprinkled with a branch of birch dipped in the blood of the victim, and then a fire of cypress and alderwood must be lighted, the two magical candles being placed on the right and left of the operator, encircled with the wreaths of vervain. The formula of evocation can now be pronounced, as they are found in the magical elements of Peter of Apono, or in the grimoires, whether printed or manuscript. That of the Grand Grimoire, reproduced in the vulgar Red Dragon, has been willfully altered, and should be read as follows:

> "By Adonaï Eloïm, Adonaï Jehova, Adonaï Sabaoth, Metraton On Agla Adonaï Mathon, the pythonic word, the mystery of the salamander, the assembly of sylphs, the grotto of gnomes, the demons of the heaven of Gad, Almousin, Gioor, Jehosua, Evam, Zariatnatmik: Come, Come, Come!"

The grand appellation of Agrippa[25] consists only in these words: DIES MIES JESCHET BOENEDOESEF DOUVEMA ENITEMAUS. We make no pretense of understanding their meaning; possibly they possess none, assuredly none which is reasonable, since they avail in evoking the devil, who is the sovereign unreason. Picus de Mirandola, no doubt from the same motive, affirms that in black magic the most barbarous and unintelligible words are the most efficacious and the best. The conjurations are repeated in a louder voice, accompanied by

---

[25] The so-called *Fourth Book of Occult Philosophy*, which was originally attributed to Agrippa, but which is now believed to be by an unknown author.

imprecations and menaces, until the spirit replies. He is commonly preceded by a violent wind which seems to make the whole country resound. Then domestic animals tremble and hide away, the assistants feel a breath upon their faces, and their hair, damp with cold sweat, rises upon their heads. The grand and supreme appellation, according to Peter of Apono, is as follows:

> *"Hemen-Etan! Hemen-Etan! Hemen-Etan!* El* Ati* Titeip* Aozia* Hyn* Teu* Minosel* Achadon* vay* vaa* Eye* Aaa* Eie* Exe* A EL EL EL A Hy! Hau! Hau! Hau! Hau! Va! Va! Va! Va! CHAVAJOTH. Aie Saraye, aie Saraye, aie Saraye! By Eloym, Archima, Rabur, Bathas over Abrac, flowing down, coming from above Abeor upon Aberer *Chavajoth Chavajoth! Chavajoth!* I command thee by the Key of Solomon and the great name Semhamphoras."

The ordinary signs and signatures of demons are given below:

But they are those of inferior demons, and here follow the official signatures of the princes of hell, attested judicially – judicially, O M. le Comte de Mirville! – and preserved in the archives of justice as convincing evidences for the trial of the unfortunate Urbain Grandier.

These signatures appear under a pact of which Collin de Plancy gives a facsimile reproduction in the Atlas of his "Infernal Dictionary." It has this marginal note: "The draught is in hell, in the secretary of Lucifer," a valuable item of information about a locality but imperfectly known, and belonging to a period approximate to our own, though anterior to the trial of young Labarre and Etalonde, who, as everyone knows, were contemporaries of Voltaire.

Evocations were followed frequently by pacts written on parchment of goat skin with an iron pen and blood drawn from the left arm. The document was in duplicate; one copy was carried off by the fiend and the other swallowed by the willful reprobate. The reciprocal engagements were that the demon should serve the sorcerer during a given period of years, and that the sorcerer should belong to the demon after a determinate time. The Church in her exorcisms has consecrated the belief in all these things, and it may be said that black magic and its darksome prince are the true, living and terrible creation of Roman Catholicism; that they are even its special and characteristic work, for priests invent not God. So, do true Catholics cleave from the bottom of their hearts to the consecration and even the regeneration of this great work, which is the philosophical stone of the official and positive cultus. In thieves' slang the devil is called *baker* by malefactors; all our desire, and we speak no longer from the standpoint of the magus, but as a devoted child of Christianity and of that Church to which we owe our earliest education and our first enthusiasms – all our desire, we say, is that the phantom of Satan may no longer be able to be termed the *baker* for the ministers of morality and representatives of the highest virtue. Will they appreciate our intention and forgive the boldness of our aspirations in consideration of our devoted intentions and the sincerity of our faith?

The devil-making magic which dictated the grimoire of Pope Honorius, the Enchiridion of Leo III, the exorcisms of the Ritual, the verdicts of inquisitors, the suits of Laubardement, the articles of the Veuillot brothers, the books of MM. de Falloux, de Montalembert, de Mirville, the magic of sorcerers and of pious persons who are not sorcerers, is something truly to be condemned in the one and infinitely deplored in the other. It is above all to combat these unhappy aberrations of the human mind by their exposure that we have published this book. May it further the holy cause!

But we have not yet exhibited these impious devices in all their turpitude, and in all their monstrous folly; we must remove the blood-stained filth of perished superstitions; we must tax the annals of demonomania, so as to conceive of certain crimes which imagination alone could not invent. The Kabbalist Bodin, Israelite by conviction and Catholic by necessity, had no other intention in his "Demonomania of Sorcerers" than to impeach Catholicism in its works, and undermine it in the greatest of all its doctrinal abuses. The treatise of Bodin is profoundly Machiavellic, and strikes at the heart of the institutions and persons it appears to defend. It would be difficult to imagine without reading his vast mass of sanguinary and hideous histories, acts of revolting superstition, sentences and executions of stupid ferocity. "Burn all!" the inquisitors seemed to cry. "God will distinguish His own!" Poor fools, hysterical women and idiots, were accordingly burned without mercy for the crime of magic, while, at the same time, great criminals escaped this unjust and sanguinary justice. Bodin gives us to understand this by recounting such anecdotes as that which he connects with the death of Charles IX. It is an almost unknown abomination, and one which has not, so far as we know, tempted the skill of any romancer, even at the periods of the most feverish and deplorable literature.

Attacked by a disease of which no physician could discover the cause or explain the frightful symptoms, King Charles IX was dying. The Queen-Mother, who ruled him entirely, and had everything to lose under another reign – the Queen-Mother, who has been suspected as author of the disease, because concealed devices and unknown interests have been always been attributed to her who was capable of anything – consulted her astrologers, and then had recourse to the foulest form of magic, the *Oracle of the Bleeding Head*, for the sufferer's condition grew worse and more desperate daily. The infernal operation was performed in the following way. A child was selected, of beautiful appearance and innocent manners; he was prepared for his first communion by the almoner of the palace. When the day or rather night of the sacrifice arrived, a monk, an apostate Jacobin, given over to the occult works of black magic, celebrated the Mass of the Devil at midnight, in the sick-room, and in the presence only of Catherine de Medicis and her trusted confidants. It was offered before the image of the demon, having a crucifix upside down under its feet, and the sorcerer consecrated two hosts, one black and one white. The white was given to the child, who was brought in clothed as for baptism, and was murdered on the steps of the altar immediately after his communion. His head, cut by one blow from the body, was set palpitating upon the great black host which covered the bottom of the paten, and then transported to a table where mysterious lamps were burning. The exorcism began, an oracle was besought of the demon, and an answer by the mouth of the head to a secret question which the king dared not make aloud, and had confided to no one. A strange and feeble voice, which had nothing human about it, was heard presently in the poor little martyr's head, saying in Latin: *Vim patior* – "I am forced." At this reply, which doubtless announced to the sick man that hell no longer protected him, a horrible trembling seized the monarch, his arms stiffened, and he cried in a hoarse voice: "Away with that head! Away with that head!" and so continued screaming till he gave up the ghost. His attendants, who were not in the confidence of the frightful mystery, believed that he was pursued by the phantom of Coligny, and that he saw the head of the illustrious admiral; what tormented the dying man was not, however, a remorse but a hopeless terror of an anticipated Hell.

This darksome magical legend of Bodin recalls the abominable practices and deserved fate of Gilles de Laval, Lord of Retz, who passed from asceticism to black magic, and offered the most revolting sacrifices to conciliate the favor of Satan. This madman confessed at his trial that Satan had frequently appeared to him, but had always deceived him by promises of treasures which he had never given. It transpired from the judicial information that several hundred unfortunate children had fallen victims to the cupidity and atrocious fancies of this monster.

Chapter XVI: Witchcraft and Spells

WHAT sorcerers and necromancers sought above all in their evocations of the impure spirit was that magnetic power which is the possession of the true adept, but was desired by them only that they might shamefully abuse it. The folly of sorcerers was an evil folly, and one of their chief ends was the power of bewitchments or harmful influences. We have set down in our Doctrine what we think upon the subject of bewitchment, and how it seems to us a dangerous and real power. The true magus bewitches without ceremonial, and by his mere reprobation, those whom he condemns and considers it necessary to punish; his forgiveness even bewitches those who do him wrong, and never do the enemies of initiates carry far the impunity of their injustice. We ourselves have witnessed numerous examples of this fatal law. The murderers of martyrs always perish miserably, and the adepts are martyrs of intelligence; Providence seems to scorn those who despise them, and to slay those who would deprive them of life. The legend of the Wandering Jew is the popular poetry of this arcanum. A wise man was driven by a nation to his doom; it bade him "Go on!" when he wished to rest for a moment. What is the consequence? A similar condemnation overtakes the nation itself; it is proscribed bodily; men have cried to it: "Get on! Get on!" for centuries, and it has found no pity and no repose.

A man of learning had a wife whom he loved wildly and passionately in the exaltation of his tenderness; he honored her with blind confidence, and trusted her entirely. Vain of her beauty and understanding, this woman became jealous of her husband's superiority, and began to hate him. Sometime after she deserted him, disgracing herself with an old, ugly, stupid and immoral man. This was the beginning of her punishment, but it did not end there. The man of learning pronounced solemnly the following sentence upon her: "I take back your understanding and your beauty." A year after she was no longer recognized by those who had known her; she had lost her plumpness, and reflected in her countenance the hideousness of her new affections. Three years later she was ugly; seven years later she was deranged. This happened in our own time, and we were acquainted with both persons.

The magus condemns, after the manner of the skillful physician, and for this reason there is no appeal from his sentence when it has once been pronounced against a guilty person. There are no ceremonies and no invocations; he simply abstains from eating at the same table, or if forced to do so, neither accepts nor offers salt. But the bewitchments of sorcerers are of another kind, and may be compared to an actual poisoning of some current of astral light. They exalt their will by ceremonies till it becomes venomous at distance; but, as we have observed in our Doctrine, they more often expose themselves to be the first that are killed by their infernal machinery. Let us here stigmatize some of their guilty proceedings. They procure hair or garments of the person whom they seek to execrate; they next select some animal, which seems to them symbolic of the person, and, by means of the hair or garments, they place it in magnetic connection with him or her. They give it the same name, and then slay it with one blow of the magic knife. They cut open the breast, tear out the heart, wrap it, while still palpitating, in the magnetized objects, and hourly, for the space of three days, they drive nails, red hot pins, or long thorns therein, pronouncing maledictions upon the name of the bewitched person. They are persuaded, and often rightly, that the victim of their infamous operations experiences as many tortures as if his own heart had been pierced at all points. He begins to waste away, and after a time dies of an unknown disease.

Another bewitchment, made use of in country places, consists in consecration of nails to works of hatred by means of the stinking fumigations of Saturn and invocations of evil genii; then, in following the footsteps of the person whom it is sought to torment, and nailing cross-wise every imprint of his feet which can be traced upon the earth or sand. Yet another and more abominable practice. A fat toad is selected; it is baptized; the name and surname of the person to be bewitched is given it; it is made to swallow a consecrated host, over which the formulæ of execration have been pronounced. The animal is then wrapped in the magnetized objects, tied with the hairs of the victim, upon which the operator has previously spat, and buried at the threshold of the bewitched person's door, or at some point where he is obliged to pass daily. The elementary spirit of the toad will become a nightmare and vampire, haunting the dreams of the victim, unless, indeed, he should know how to send it back to the operator.

Let us pass now to bewitchments by waxen images. The sorcerers of the middle ages, eager to please by their sacrileges him whom they regarded as their master, mixed baptismal oil and ashes of consecrated hosts with a modicum of wax. Apostate priests were never wanting to deliver them the treasures of the Church. With the accursed wax they formed an image as much as possible resembling the person whom they desired to bewitch. They clothed this image with garments similar to his, they administered to it the sacraments which he received, then they called down upon its head all maledictions which could express the hatred of the sorcerer, inflicting daily imaginary tortures upon it, so as to reach and torment by sympathy the person represented by the image. This bewitchment is more infallible if the hair, blood, and, above all, a tooth of the victim can be procured. It was this which gave rise to the proverbial saying: You have a tooth against me. There is also bewitchment by the glance, called the *jettatura*, or evil eye, in Italy. During our civil wars, a shopkeeper had the misfortune to denounce one of his neighbors, who, after a period of detention, was set at liberty, but with his position lost. His sole vengeance was to pass twice daily the shop of his denouncer, whom he regarded fixedly, saluted, and went on. Some little time after, the shopkeeper, unable to bear the torment of this glance any longer, sold his goods at a loss, and changed his neighborhood, leaving no address. In a word, he was ruined.

A threat is a real bewitchment, because it acts powerfully on the imagination, above all, when the latter receives with facility the belief in an occult and unlimited power. The terrible menace of hell, that bewitchment of humanity during so many centuries, has created more nightmares, more nameless diseases, more furious madness, than all vices and all excesses combined. This is what the Hermetic artists of the middle ages represented by the incredible and unheard-of monsters which they carved at the doors of basilicas. But bewitchment by threat produces an effect altogether contrary to the intentions of the operator when it is evidently a vain threat, when it does outrage to the legitimate pride of the menaced person, and consequently provokes his resistance, or, finally, when it is ridiculous by its atrocity. The sectaries of hell have discredited heaven. Say to a reasonable man that equilibrium is the law of motion and life, and that liberty, which is moral equilibrium, rests upon an eternal and immutable distinction between true and false, between good and bad; tell him that, endowed as he is with free will, he must place himself by his works in the empire of truth and goodness, or relapse eternally, like the rock Sisyphus, into the chaos of falsehood and evil; then he will understand the doctrine, and if you term truth and goodness heaven, falsehood and evil hell, he will believe in your heaven and hell, over which the divine ideal rests calm, perfect and inaccessible to either wrath or offence, because he will understand that if in principle hell be eternal as liberty, it cannot in fact be more than a temporary agony for souls, because it is an expiation, and the idea of expiation necessarily supposes that of reparation and destruction of evil. This much said, not with dogmatic intention, which is outside our province, but to indicate the moral and reasonable remedy for the bewitchment of consciences by the terrors of the life beyond, let us speak of the means of escaping the baleful influences of human wrath. The first among all is to be reasonable and just, giving no opportunity or excuse to anger. A lawful indignation is greatly to be feared; make haste therefore to acknowledge and expiate your faults. Should anger persist after that, then it certainly proceeds from vice; seek to know what vice, and unite yourself strongly to the magnetic currents of the opposite virtue. The bewitchment will then have no further power upon you. Wash carefully the clothes which you have finished with before giving them away; otherwise, burn them; never use a garment which has belonged to an unknown person without purifying it by water, sulphur, and such aromatics as camphor, incense, amber, etc.

A great means of resisting bewitchment is not to fear it; it acts after the manner of contagious maladies. In times of epidemic, the terror-struck are the first to be attacked. The secret of not fearing an evil is not to think about it, and my advice is completely disinterested since I give it in a book on magic of which I am the author, when I strongly urge upon persons who are nervous, feeble, credulous, hysterical, superstitious devotees, foolish, without energy and without will, never to open a book on magic, to close this one if they have opened it, to turn a deaf ear to those who talk of the occult sciences, to deride them, never to believe in them, and to drink water, as said the great pantagruelist magician, the excellent curé of Meudon.

As for the wise – and it is time that we turned to them after espousing the cause of the foolish – they have scarcely any sorceries to fear save those of fortune, but seeing that they are priests and physicians, they may be called upon to cure the bewitched, and this should be their method of procedure. They must persuade a bewitched person to do some act of goodness to his bewitcher, render him some service which he cannot refuse, and lead him directly or otherwise to the communion of salt. A person who believes himself bewitched by the execration and interment of the toad must carry about him a living toad in a horn box. For the bewitchment of the pierced heart, the afflicted individual must be made to eat a lamb's heart seasoned with sage and onion, and to carry a talisman of Venus or of the moon in a satchel filled with camphor and salt. For bewitchment by the waxen figure, a more perfect figure must be made, as much as possible in the likeness of the person; seven talismans must be hung round the neck; it must be set in the middle of a great pantacle representing the pentagram, and must each day be rubbed slightly with a mixture of oil and balm, after reciting the Conjuration of the Four to turn aside the influence of elementary spirits. At the end of seven days the image must be burnt in consecrated fire, and one may rest assured that the figure fabricated by that bewitcher will at such moment lose all its virtue.

We have already mentioned the sympathetic medicine of Paracelsus, who medicated waxen limbs and operated upon discharges of blood from wounds for the cure of the wounds themselves. This system permitted the employment of more than usually violent remedies, and his chief specifics were sublimate and vitriol. We believe that homœopathy is a reminiscence of the theories of Paracelsus and a return to his wise practices. But we shall follow up this subject in a special treatise exclusively consecrated to occult medicine.

Contracts by parents forestalling the future of their children are bewitchments which cannot be too strongly condemned; children dedicated in white, for example, scarcely ever prosper; those who were formerly devoted to celibacy fell commonly into debauch, or ended in despair and madness. Man is not permitted to do violence to destiny, still less to impose bonds upon the lawful use of liberty.

As a supplement or appendix to this chapter, we will add a few words about mandragores and androids, which several writers on magic confound with the waxen images serving the purposes of bewitchment. The natural mandragore is a filamentous root which, more or less, presents as a whole either the figure of a man, or that of the virile members. It is slightly narcotic, and an aphrodisiacal virtue was ascribed to it by the ancients, who represented it as being sought by Thessalian sorcerers for the composition of philtres. Is this root the umbilical vestige of our terrestrial origin? We dare not seriously affirm it, but all the same it is certain that man came out of the slime of the earth, and his first appearance must have been in the form of a rough sketch. The analogies of nature make this notion necessarily admissible, at least as a possibility. The first men were, in this case, a family of gigantic, sensitive mandragores, animated by the sun, who rooted themselves up from the earth; this assumption not only does not exclude, but, on the contrary, positively supposes, creative will and the providential co-operation of a first cause, which we have REASON to call GOD.

Some alchemists, impressed by this idea, speculated on the culture of the mandragore, and experimented in the artificial reproduction of a soil sufficiently fruitful and a sun sufficiently active to humanize the said root, and thus create men without the concurrence of the female. Others, who regarded humanity as the synthesis of animals, despaired about vitalizing the mandragore, but they crossed monstrous pairs and projected human seed into animal earth, only for the production of shameful crimes and barren deformities. The third method of making the android was by galvanic machinery. One of these almost intelligent automata was attributed to Albertus Magnus, and it is said that St. Thomas destroyed it with one blow from a stick because he was perplexed by its answers. The story is an allegory; the android was primitive scholasticism, which was broken by the *Summa* of St. Thomas, the daring innovator who first substituted the absolute law of reason for arbitrary divinity, by formulating that axiom which we cannot repeat too often, since it comes from such a master: "A thing is not just because God wills it, but God wills it because it is just."

The real and serious android of the ancients was a secret which they kept hidden from all eyes, and Mesmer was the first who dared to divulge it; it was the extension of the will of the magus into another body, organized and served by an elementary spirit; in more modern and intelligible terms, it was a magnetic subject.

Chapter XVII: The Writing of the Stars

WE have finished with infernus, and we breathe the fresh air freely as we turn to daylight after traversing the crypts of black magic. Get thee behind us, Satan! We renounce thee, with all thy pomps and works, and still more with all thy deformities, thy meanness, thy nothingness, thy deception! The Great Initiator beheld thee fall from heaven like a thunderbolt. The Christian legend changes thee, making thee set thy dragon's head mildly beneath the foot of the mother of God. Thou art for us the image of unintelligence and mystery; thou art unreason and blind fanaticism; thou art the inquisition and its hell; thou art the god of Torquemada and Alexander VI; thou hast become the sport of children, and thy final place is at the side of Polichinello; henceforth thou art only a grotesque figure in our foreign theatres, and a means of instruction in a few so-called religious markets.

After the sixteenth Key of the Tarot, which represents the downfall of Satan's temple, we find on the seventeenth leaf a magnificent and gracious emblem. A naked woman, a young and immortal maid, pours out upon the earth the juice of universal life from two ewers, one of gold and one of silver; hard by there is a flowering shrub, on which rests the butterfly of Psyche; above shines an eight-pointed star with seven other stars around it. "I believe in eternal life!" Such is the final article of the Christian symbol, and this of itself is a full profession of

Transcendent Magic

faith.

The ancients, when they compared the calm and peaceful immensity of heaven, thronged with innumerable lights, to the tumults and darkness of this world, believed themselves to have discovered in that beautiful book, written in letters of gold, the final utterance of the enigma of destinies; in imagination they drew lines of correspondence between these shining points of the divine writing, and it is said that the first constellations marked out by the shepherds of Chaldea were also the first letters of the kabbalistic alphabet. These characters, expressed first of all by means of lines, then enclosed in hieroglyphic figures, would, according to M. Moreau de Dammartin, author of a very curious treatise on alphabetic characters, have determined the ancient magi in the choice of the Tarot figures, which are considered by this man of learning, as by ourselves, for an essentially hieratic and primitive book. Thus, in his opinion, the Chinese *tseu*, the Hebrew *aleph*, and the Greek *alpha*, expressed hieroglyphically by the figure of the juggler, would be borrowed from the constellation of the crane, in the vicinity of the celestial fish, a sign of the eastern hemisphere. The Chinese *tcheou*, the Hebrew *beth* and the Latin B, corresponding to Pope Joan or Juno, were formed after the head of the ram; the Chinese *yn*, the Hebrew *ghimel* and the Latin G, represented by the Empress, would be derived from the constellation of the Great Bear, etc. The kabbalist Gaffarel, whom we have cited more than once, erected a planisphere, in which all the constellations form Hebrew letters; but we confess that the configurations are frequently arbitrary in the highest degree, and upon the indication of a single star, for example, we can see no reason why a ד should be traced rather than a ו or ן; four stars will also give indifferently a ה, ח, or ח, as well as an א.

We are therefore deterred from reproducing a copy of Gaffarel's planisphere, examples of which are, moreover, not exceedingly rare. It was included in the work of Montfauçon on the religions and superstitions of the world, and also in the treatise upon magic published by the mystic Eckartshausen. Scholars, moreover, are unagreed upon the configuration of the letters of the primitive alphabet. The Italian Tarot, of which the lost Gothic originals are much to be regretted, connects by the disposition of its figures with the Hebrew alphabet in use after the captivity, and known as the Assyrian alphabet; but there are fragments of anterior Tarots where the disposition is different. There should be no conjecture in matters of research, and hence we suspend our judgement in the expectation of fresh and more conclusive discoveries. As to the alphabet of the stars, we believe it to be intuitive, like the configuration of clouds, which seem to assume any form that imagination lends them. Star-groups are like points in geomancy or the pasteboards of cartomancy. They are a pretext for self-magnetizing, an instrument to fix and determine native intuition. Thus, a kabbalist, familiar with mystic hieroglyphics, will perceive signs in the stars which will not be discerned by a simple shepherd, but the shepherd, on his part, will observe combinations that will escape the kabbalist. Country people substitute a rake for the belt and sword of Orion, while a kabbalist recognizes in the same sign as a whole all the mysteries of Ezekiel, the ten sephiroth arranged in a triadic manner, a central triangle formed of four stars, then a line of three stars making the jod, and the two figures taken together expressing the mysteries of Bereschith; finally, four stars constituting the wheels of Mercavah, and completing the divine chariot. Looked at after another manner, and arranging other ideal lines, he will notice a well-formed *ghimel* placed above a *jod*, in a large *daleth*, a symbol typifying the strife between good and evil, with the final triumph of good. As a fact, the *ghimel* superposed on the *jod* is the triad produced by unity, the manifestation of the divine Word, whilst the reversed *daleth* is the triad composed of the evil duad multiplied by itself.

Thus regarded, the figure of Orion would be identical with that of the angel Michael doing battle with the dragon, and the appearance of this sign, so understood, would be, for the kabbalist, a portent of victory and happiness.

A long contemplation of the sky exalts imagination, and then the stars respond to our thoughts. The lines drawn mentally from one to another by primitive observers have given man his first notions of geometry. Accordingly, as our soul is troubled or at rest, the stars seem burning with menace or sparkling with hope. The sky is thus the mirror of the human soul, and when we think that we are reading in the stars it is in ourselves we read.

Gaffarel, applying the prophesies of celestial writing to the destinies of empires, says that not in vain did the ancients place all signs of evil augury in the northern region of the sky; calamities have been in all ages regarded as coming from the north to spread themselves over the earth by the invasion of the south. "For this reason," he tells us, "the ancients represented in the northern parts of the heaven a serpent or dragon near two bears, since these animals are the true hieroglyphs of tyranny, pillage and all oppression. As a fact, glance at history, and you will see that all great devastations proceed from the north. The Assyrians or Chaldeans, incited by Nebuchodonosor or Salmanasor, exhibited this truth in abundance by the destruction of the most splendid and most holy temple and city in the universe, and by the complete overthrow of a people whom God himself had taken under his special protection, of whom he specially termed himself father. So also that other Jerusalem, Rome the blessed, has it not too experienced frequently the violence of this evil northern race, when it beheld its altars demolished and the towers of its proud edifices brought level with the foundations, through the cruelty of Alaric, Genseric, Attila, and other princes of the Goths, Huns, Vandals and Alain . . . Very properly, therefore, in the secrets of this celestial writing, do we read calamities and misfortunes on the northern side, since a *septentrione pandetur omne malum*. Now, the word הפתח , which we translate by *pandetur*, is also an equivalent of *depingetur* or *scribetur*, and the prophecy signifies equally: All misfortunes of the world are written in the northern sky."

We have transcribed this passage at length, because it is not without application in our own day, when the north once more seems to threaten Europe;[26] but it is also the destiny of hoar-frost to be melted by the sun, and the darkness disappears of itself when the light manifests. Such is our final word of prophecy, and the secret of the future. Gaffarel adds some prognostics drawn from the stars, as, for example, the progressive weakening of the Ottoman empire; but, as already said, his constellated letters are exceedingly arbitrary. He states, for the rest, that he derived his predictions from a Hebrew kabbalist, Rabbi Chomer, but does not himself pretend to understand him especially well.

Here follows a table of magical characters traced after the zodiacal constellations by the ancient astrologers; each of them represents the name of a genius, be he good or evil. It is known that the Signs of the Zodiac correspond to various celestial influences and consequently signify an annual alternative of good or evil.

---

[26] This passage was written before the Crimean War.

Éliphas Lévi

The names of the genii designated by the above characters are: For the Ram, SATAARAN and *Sarahiel*; for the Bull, BAGDAL, and *Araziel*; for the Twins, SAGRAS and *Saraïel*; for the Crab, RAHDAR and *Phakiel*; for the Lion, SAGHAM and *Seratiel*; for the Virgin, IADARA and *Schaltiel*; for the Balance, GRASGARBEN and *Hadakiel*; for the Scorpion, RIEHOL and *Saissaiel*; for the Archer, VHNORI and *Saritaïel*; for the Goat, SAGDALON and *Semakiel*; for the Water-Bearer, ARCHER and *Ssakmakiel*; for the Fishes, RASAMASA and *Vacabiel*.

The wise man, who would read the sky, must observe also the days of the moon, the influence of which is very great in astrology. The moon successively attracts and repels the magnetic fluid of the earth, and thus produces the ebb and flow of the sea; we must, therefore, be well acquainted with its phases and be able to distinguish its days and hours. The new moon is propitious to the beginning of all magical works: from first quarter to full moon its influence is warm; from full moon to third quarter it is dry; and from third quarter to last it is cold. Here follow the special characters of all days of the moon, distinguished by the twenty-two Tarot keys and by the signs of the seven planets.

1. *The Juggler, or Magus.*

The first day of the moon is that of the creation of the moon itself. This day is consecrated to mental enterprises, and should be favorable for opportune innovations.

2. *Pope Joan, or Occult Science.*

The second day, the genius of which is Enediel, was the fifth of creation, for the moon was made on the fourth day. The birds and fishes, created on this day, are living hieroglyphs of magical analogies and of the universal doctrine of Hermes. The water and air, which were thereby filled with forms of the Word, are elementary figures of the Mercury of the Sages, that is, of intelligence and speech. This day is propitious to revelations, initiations, and great discoveries of science.

3. *The Celestial Mother, or Empress.*

The third day was that of man's creation. So is the moon called the MOTHER in Kabbalah, when it is represented in association with the number 3. This day is favorable to generation, and generally to all productions, whether of body or mind.

4. *The Emperor, or Ruler.*

The fourth day is baleful; it was that of the birth of Cain; but it is favorable to unjust and tyrannical enterprises.

5. *The Pope, or Hierophant.*

The fifth day is fortunate; it was that of the birth of Abel.

6. *The Lover, or Liberty.*

The sixth day is a day of pride; it was that of the birth of Lamech, who said unto his wives: "I have slain a man to my wounding, and a young man to my hurt. If Cain shall be avenged sevenfold, truly Lamech seventy and sevenfold." This day is propitious for conspiracies and rebellions.

7. *The Chariot.*

On the seventh day, birth of Hebron, who gave his name to the first of the seven sacred cities of Israel. A day of religion, prayers and success.

8. *Justice.*

Murder of Abel. Day of expiation.

9. *The Old Man, or Hermit.*

Birth of Methuselah. Day of blessing for children.

10. *Ezekiel's Wheel of Fortune.*

Birth of Nebuchodonosor. Reign of the Beast. Fatal day.

11. *Strength.*

Birth of Noah. Visions on this day are deceitful, but it is one of health and long life for children born on it.

12. *The Victim, or Hanged Man.*

Birth of Samuel. Prophetic and kabbalistic day, favorable to the fulfilment of the great work.

13.   *Death.*

Birthday of Canaan, the accursed son of Cham. Baleful day and fatal number.

14.   *The Angel of Temperance.*

Blessing of Noah on the fourteenth day of the moon. This day is governed by the angel Cassiel of the hierarchy of Uriel.

15.   *Typhon, or the Devil.*

Birth of Ishmael. Day of reprobation and exile.

16.   *The Blasted Tower.*

Birthday of Jacob and Esau; the day also of Jacob's predestination, to Esau's ruin.

17.   *The Glittering Star.*

Fire from heaven burns Sodom and Gomorrah. Day of salvation for the good, and ruin for the wicked; on a Saturday dangerous. It is under the dominion of the Scorpion.

18.   *The Moon.*

Birth of Isaac. Wife's triumph. Day of conjugal affection and good hope.

19.   *The Sun.*

Birth of Pharoah. A beneficent or fatal day for the great of earth, according to the different merits of the great.

20. *The Judgement.*

Birth of Jonah, the instrument of God's judgement. Propitious for divine revelations.

21. *The World.*

Birth of Saul, material royalty. Danger to mind and reason.

22. *Influence of Saturn.*

Birth of Job. Day of trial and suffering.

23. *Influence of Venus.*

Birth of Benjamin. Day of preference and tenderness.

24. *Influence of Jupiter.*

Birth of Japhet.

25. *Influence of Mercury.*

Tenth plague of Egypt.

26. *Influence of Mars.*

Deliverance of the Israelites, and passage of the Red Sea.

27. *Influence of Diana, or Hecate.*

Splendid victory achieved by Judas Maccabeus.

28. *Influence of the Sun.*

Samson carries off the gates of Gaza. Day of strength and deliverance.

          29.   *The Fool of the Tarot.*

Day of failure and miscarriage in all things.

We see from this rabbinical table, which John Belot and others borrowed from the Hebrew kabbalists, that these ancient masters concluded *à posteriori* from facts to presumable influences, which is completely within the logic of the occult sciences. We see also what diverse significations are included in the twenty-two keys which form the universal alphabet of the Tarot, together with the truth of our affirmation, when we say that all secrets of the Kabbalah and magic, all mysteries of the elder world, all science of the patriarchs, all historical traditions of primeval times, are enclosed in this hieroglyphic book of Thoth, Enoch or Cadmus.

An exceedingly simple method of finding celestial horoscopes by onomancy is that which we are about to describe; it harmonizes Gaffarel with our own views, and its results are most astounding in their exactitude and depth. Take a black card; cut therein the name of the person for whom you wish to make the consultation; place this card at the end of a tube which must diminish towards the eye of the observer; then look through it alternately towards the four cardinal points, beginning at the east and finishing at the north. Take note of all the stars which you see through the letters; convert these letters into numbers, and, with the sum of the addition written down in the same manner, renew the operation; then compute the number of stars you have; next, adding this number to that of the name, again cast up and write the sum of the two numbers in Hebrew characters. Again, renew the operation; inscribe separately the stars which you have noticed; then find the names of all the stars in the planisphere; classify them according to their size and brightness, choosing the most brilliant of all as the pole-star of your astrological operation; then find, in the Egyptian planisphere, the names and figures of the genii to which these stars belong. A good example of the planisphere will be found in the atlas to the great work of Dupuis. You will then know the fortunate and unfortunate signs which enter into the name of the person, and what is their influence; whether in childhood, which is the name traced at the east; in youth, which is the name traced at the south; in mature age, which is the name at the west; in decline, which is the name at the north; or, finally, during the whole life, obtained from the stars which enter into the entire number formed by the addition of letters and stars. This astrological operation is simple, easy, and requires few calculations; it connects with the highest antiquity, and belongs evidently to primitive patriarchal magic, as will be seen by studying the works of Gaffarel and his master Rabbi Chomer. Onomantic astrology was practiced by the old Hebrew kabbalists, as is proved from their observations by Rabbi Chomer, Rabbi Kapol, Rabbi Abjudan, and other masters in Kabbalah. The menaces of the prophets uttered against various nations were based upon the characters of the stars found vertically over them in the permanent correspondence of the celestial and terrestrial spheres. Thus, by writing in the sky of Greece the Hebrew name of that country יון or יוג, and translating in numbers, they obtained the word חרב, which signifies destroyed, desolated.

ח ר ב
2 2 8

CHARAB.

Destroyed, Desolated.

Sum 12.

י ו נ
5 6 1

JAVAN.

Greece.

Sum 12.

Hence, they inferred that after a cycle of twelve periods Greece would be destroyed and desolated. A short time before the sack of Jerusalem and its temple by Nebuzardan, the kabbalists remarked eleven stars disposed in the following manner vertically above the temple:

```
*   *   *   *   *   *   *   *
            *
        *       *
```

All these entered into the word הבשיה written from south to west, the term signifying reprobation and abandonment without mercy. The sum of the number of the letters is 423, exactly the period of the duration of the temple. Destruction threatened the empires of Persia and Assyria, in the shape of four vertical stars which entered into the three letters רוב, *Roev*, and the fatal period indicated was 208 years. So, also, four stars announced to the kabbalistic rabbins of another period the fall and division of the empire of Alexander; they entered into the word פרד, *Parad*, to divide, 284, the number of this word, indicating the entire duration of this empire, both as to root and branches. According to Rabbi Chomer, the destinies of the Ottoman power at Constantinople would be fixed and foretold by four stars, entering into the word באה, *Caah*, signifying to be feeble, weak and drawing to its end. The stars being more brilliant in the letter א Aleph, indicated a capital, and gave it the numerical value of a thousand. The three letters combined make 1025, which must be computed from the taking of Constantinople by Mahomet II, a calculation which still promises several centuries of existence to the enfeebled empire of the sultans, at present sustained by all Europe combined. The MANE THECEL PHARES which Balthazar, in his intoxication, saw written on the wall of his palace by the glare of torches, was an onomantic intuition similar to that of the rabbins. Initiated, no doubt, by his Hebrew diviners in the reading of the stars, Balthazar operated mechanically and instinctively upon the lamps of his nocturnal feast, as he would upon the stars of heaven. The three words which he had formed in his imagination soon became indelible to his eyes, and paled all the lights of his banquet. It was easy to predict an end like that of Sardanapalus to a king who abandoned himself to orgies in a besieged town.

In conclusion, we have said and we repeat that magnetic intuitions alone give value and reality to these kabbalistic and astrological calculations, puerile possibly, and completely arbitrary, when made without inspiration, by cold curiosity, and in the absence of a powerful will.

Chapter XVIII: Philtres and Magnetism

LET us now adventure in Thessaly, the country of enchantments. Here was Apuleius deluded like the companions of Ulysses, and underwent a shameful metamorphosis. Here all is magical, – the birds that fly, the insects humming in the grass, even the trees and flowers; here in the moonlight are brewed those potions which compel love; here spells are devised by stryges to render them young and lovely like the Charites. O all ye youths, beware!

The art of poisoning the reason, or of philtres, seems, as a fact, if traditions may be trusted, to have developed its venomous efflorescence more abundantly in Thessaly than elsewhere; there, also, magnetism played its most important part, for exciting or narcotic plants, bewitching and harmful animal substances, derived all their power from enchantments – that is to say, sacrifices accomplished and words pronounced by sorcerers when preparing philtres and beverages. Stimulating substances, and those in which phosphorus predominates, are naturally aphrodisiacal. Anything which acts strongly on the nervous system may determine impassioned exaltation, and when a skillful and persevering will knows how to direct and influence these natural tendencies, it can make use of the passions of others to the profit of its own, and will soon reduce the most independent personalities into instruments of its pleasures. From such influence it behooves us to seek protection, and to give arms to the weak is our purpose in writing this chapter. These, in the first place, are the devices of the enemy. The man who seeks to compel love – we attribute such unlawful manœuvres to men only, assuming that women can never have need of them – must in the first place make himself observed by the person whom he desires, and must contrive to impress her imagination. He must inspire her with admiration, astonishment, terror, even with horror, failing all other resources; but at any cost he must set himself apart in her eyes from the rank of ordinary men and, with or against her will, must make himself a place in her memory, her apprehensions, her dreams. The type of Lovelace is certainly not the admitted ideal of the type of Clarissa, but she thinks of him incessantly to condemn him, to execrate him, to compassionate his victims, to desire his conversion and repentance; next she seeks his regeneration by devotion and forgiveness; later on, secret vanity whispers to her how grand it would be to fix the affections of a Lovelace, to love him and yet to withstand him. Behold, then, Clarissa surprised into loving Lovelace! She chides herself, blushes, renounces a thousand times and loves him a thousand more; then, at the supreme moment, she forgets to resist him. Had angels been women, as represented by modern mysticism, Jehovah, indeed, would have acted as a wise and prudent Father by placing Satan at the gate of heaven. It is a serious imposition on the self-love of some amiable women to find that man fundamentally good and honorable who enamored them when they thought him a scapegrace. The angel leaves him disdainfully, saying: "You are not the devil!" Play the devil as well as you can, if you wish to allure an angel. No license is possible to a virtuous man. "For what does he take us?" say the women. "Does he think us less strict than he is?" But everything is forgiven in a rascal.

"What else could you expect?" The part of a man with high principles and of rigid character can never be a power save with women whom one wishes to fascinate; the rest, without exception, adore reprobates. It is quite the opposite with men, and this contrast has made modesty woman's dower, the first and most natural of her coquetries. One of the distinguished physicians and most amiable men of learning in London told me last year that one of his clients, when leaving the house of a distinguished lady, observed to him: "I have just had a strange compliment from the Marchioness of —-. Looking me straight in the face, she said: 'Sir, you will not make me flinch before your terrible glance: you have the eyes of Satan.'" "Well," answered the doctor, smiling, "you, of course, put your arms round her neck and embraced her?" "Not at all; I was overwhelmed by her sudden onslaught." "Beware how you call on her again, then, my friend; you will have fallen deeply in her estimation!"

The office of executioner is commonly said to go down from father to son. Do executioners really have children? Undoubtedly, as they never fail to get wives. Marat had a mistress who loved him tenderly, he, the loathsome leper; but still it was that terrible Marat, who caused the world to tremble. Love, above all in a woman, may be termed a veritable hallucination; for want of a prudent motive, it will frequently select an absurd one. Deceive Joconde for a baboon, what horror! – Ah! but supposing it is a horror, why not perpetrate it? It must be pleasant to be occasionally guilty of a small abomination!

Given this transcendental knowledge of the woman, another device can be adopted to attract her notice – not to concern oneself with her, or to do so in a way which is humiliating to her self-love, treating her as a child and deriding all notion of paying court to her. The parts are then reversed; she will move heaven and earth to tempt you; she will initiate you into secrets which women keep back; she will vest and unvest before you, making such observations as: "Between women–among old friends–I have no fear about you–you are not a man for me." etc. Then she will watch your expression; if she find it calm and indifferent, she will be indignant; she will approach you under some pretext, brush you with her tresses, permit her bodice to slip open. Women, in such cases, occasionally will risk a violence, not out of desire, but from curiosity, from impatience, and from provocation. A magician of any spirit will need no other philtres than these; he will also use flattering words, magnetic breathings, slight but voluptuous contacts, by a kind of hypocrisy, and as if unconscious. Those who resort to potions are old, idiotic, ugly, impotent. Where, indeed, is the use of the philtre? Anyone who is truly a man has always at his disposal the means of making himself loved, providing that he does not seek to usurp a place which is occupied. It would be a sovereign blunder to attempt the conquest of a young and affectionate bride during the first felicities of the honeymoon, or of a fortified Clarissa already made miserable by a Lovelace, or bitterly lamenting her love.

We shall not discuss here the impurities of black magic on the subject of philtres; we have done with the coctions of Canidia. The epodes of Horace tell us after what manner this abominable Roman sorceress compounded her poisons, while for the sacrifices and enchantments of love, we may refer to the Eclogues of Virgil and Theocritus, where the ceremonials for this species of magical work are minutely described. Nor shall we need to reproduce the recipes of the Grimoires or of the Little Albert, which anyone can consult for themselves. All these various practices connect with magnetism or poisonous magic, and are either foolish or criminal. Potions which enfeeble mind and disturb reason assure the empire already conquered by an evil will, and it was thus that the empress Casonia is said to have fixed the savage love of Caligula. Prussic acid is the most terrible agent in these envenomings of thought; hence we should beware of all extractions with an almond flavor, and never tolerate in bedchambers the presence of laurel-almond, *datura stramonium*, almond-soaps or washes, and generally all perfumes in which this odor predominates, above when its action on the brain is seconded by that of amber.

To weaken the activity of intelligence is to strengthen proportionally the forces of unreasoning passion. Love of that kind which the malefactors we are concerned with would inspire is a veritable stupefaction and the most shameful of moral bondages. The more we enervate a slave, the more incapable we make him of freedom, and here lies the true secret of the sorceress in Apuleius and the potions of Circe. The use of tobacco, by smoking or otherwise, is a dangerous auxiliary of stupefying philtres and brain poisons. Nicotine, as we know, is not less deadly than prussic acid and is present in tobacco in larger quantities than is this acid in almonds. The absorption of one will by another frequently changes a whole series of destinies, and not for ourselves only should we watch our relations, learning to distinguish pure from impure atmospheres, for the true philtres, and those most dangerous, are invisible; these are the currents of vital radiating light, which, mingling and interchanging, produce attractions and sympathies, as magnetic experiments leave no room to doubt. The history of the Church tells us that an arch-heretic named Marcos infatuated all women by breathing on them, but his power was destroyed when a valiant Christian female, who forestalled him in breathing, and said to him: "May God judge thee!" The curé Gaufridy, who was burnt as a sorcerer, pretended to enamor all women who came in contact with his breath. The notorious Father Girard, a Jesuit, was accused by his penitent, Mlle. Cardier, of completely destroying her self-control by breathing on her. The excuse was most necessary to minimize the horrible and ridiculous nature of her accusations against this priest, whose guilt, moreover, has never been well established, though, consciously or unconsciously, he had certainly inspired an exceedingly shameful passion in the miserable girl.

"Mlle. Ranfaing, having become a widow in 16—," says Dom Calmet in his "Treatise on Apparitions," "was sought in marriage by a physician named Poirot. Failing to obtain a hearing, he thereupon gave her potions to induce love, and these caused extraordinary derangements in the health of the lady, increasing to such a degree that she was believed to be possessed, and physicians, baffled by her case, recommended her for the exorcisms of the Church. Thereupon, by command of M. de Porcelets, Bishop of Toul, the following were named as her exorcists: M. Viardin, doctor in theology, the state councilor of the Duke of Lorraine, a Jesuit, and a capuchin, but in the long course of these ceremonies, almost all the clergy of Nancy, the aforesaid lord bishop, the bishop of Tripoli, suffragan of Strasbourg, M. de Nancy, formerly ambassador of the most Christian King at Constantinople and then priest of the Oratory, Charles of Lorraine, Bishop of Verdun, two Sorbonne doctors specially deputed to assist, exorcised her in Hebrew, in Greek and in Latin, and she invariably replied to them pertinently, though she herself could scarcely read even Latin. Mention is made of the certificate given by M. Nicholas de Harlay, learned in the Hebrew tongue, who recognized that Mlle. Ranfaing was really possessed, that she had answered the mere motion of his lips without any uttered words, and had given numerous other proofs. The sieur Gamier, doctor of the Sorbonne, having also commanded her several times in the Hebrew language, she replied lucidly, but in French, saying that the pact bound her to speak an ordinary language. The demon added: 'Is it not sufficient for me to show that I understand what you say?' The same doctor, addressing him in Greek, inadvertently used one case for another, whereupon the possessed woman, or rather the devil, said: 'You have blundered.' The doctor replied in Greek, 'Point out my error.' The devil answered, 'Be satisfied that I mention the mistake; I shall tell you no more.' The doctor bade him be silent in Greek, and he retorted, "You bid me be silent, and I will not be silent.'"

This remarkable example of hysterical affection carried into the region of ecstasy and demonomania, as the consequence of a potion administered by a man who believed that he was a sorcerer, proves better than anything we could say, the omnipotence of will and imagination reacting one upon another, and the strange lucidity of ecstatics or somnambulists, who comprehend speech by reading it in thought, though they have no knowledge of the words. I make no question as to the sincerity of the witnesses cited by Dom Calmet; I am merely astonished that men so serious failed to notice the pretended demon's difficulty over answering in a tongue foreign to the sufferer. Had their interlocutor been what they meant by a demon, he would have spoken as well as understood Greek; the one would have been as easy as the other to a spirit so acute and learned. Dom Calmet does not stop here with his history; he enumerates a long series of insidious questions and unserious injunctions on the part of the exorcisers, and a like sequence of more or less congruous replies by the poor sufferer, still ecstatic and somnambulistic. It is needless to add that the excellent father draws precisely the luminous conclusions of the not less excellent M. de Mirville. The phenomena being above the comprehension of the witnesses, they were all ascribed to perdition. Splendid and instructed conclusion! The most serious part of the business is that the physician Poirot was arraigned as a magician, confessed like all others under torture, and was burnt. Had he, by any potion, really attempted the reason of the woman in question, he would have deserved punishment as a poisoner; this is the most that we can say.

But the most terrific of all philtres are the mystical exaltations of misdirected devotion. Will ever any impurities equal the nightmares of St. Anthony or the tortures of St. Theresa and St. Angela de Foligny? The last applied a red-hot iron to her rebellious flesh, and found that the material fire was cooling to her hidden ardors. With what violence does nature cry out for that which is denied her, but is brooded over continually to increase detestation thereof! The pretended bewitchments of Magdalen Bavan, of Mlles. de la Palud and de la Cadière, began with mysticism. The excessive fear of a given thing makes it almost invariably inevitable. To follow the two curves of a circle is to reach and to meet at the same point. Nicholas Remigius, criminal judge of Lorraine, who burnt alive eight hundred women as sorcerers, beheld magic everywhere; it was his fixed idea, his mania. He was eager to preach a crusade against sorcerers, with whom Europe, in his opinion, was swarming; in despair that his word was not taken when he affirmed that nearly everyone in the world had been guilty of magic, he ended by declaring that he was himself a sorcerer, and was burned on his own confession.

To preserve ourselves against evil influences, the first condition is therefore to forbid excitement to the imagination. All those who are prone to excitement are more or less mad, and a maniac is ever governed by his mania. Place yourself, then, above puerile fears and vague desires; believe in supreme wisdom, and be assured that this wisdom, having given you understanding as the means of knowledge, cannot seek to lay snares for your intelligence or reason. Everywhere about you, you behold effects proportioned to their causes; you find causes directed and modified in the domain of humanity by understanding; in a word, you will find goodness stronger and more respected than evil; why should you assume an immense unreason in the infinite, seeing that there is reason in the finite? Truth is hidden from no one. God is visible in His works, and He requires nothing contrary to its nature from any being, for He is himself the author of that nature. Faith is confidence; have confidence, not in men who malign reason, for they are fools or impostors, but in that eternal reason which is the Divine Word, that true light which is offered like the sun to the intuition of every human creature coming into this world. If you believe in absolute reason, and if you desire truth and justice before all things, you will have no occasion to fear anyone, and you will love those only who are deserving of love. Your natural light will repel instinctively that of the wicked, because it will be ruled by your will. Thus, even poisonous substances, which it is possible may be administered to you, will not affect your intelligence; ill, indeed, they may make you, but never criminal.

What most contributes to render women hysterical is their soft and hypocritical education; if they took more exercise, if they were instructed more frankly and fully in matters of the world, they would be less capricious, and consequently less accessible to evil tendencies. Weakness ever sympathizes with vice, because vice is a weakness which assumes the mask of strength. Madness holds reason in honor, and on all subjects, it delights in the exaggerations of falsehood. In the first place, therefore, cure your diseased intelligence. The cause of all bewitchments, the poison of all philtres, the power of all sorcerers, are there. As to narcotics or other drugs which may be administered to you, it is a subject for the physician and the law, but we do not think such enormities will be largely reproduced at this day. Lovelaces no longer stupefy Clarissas otherwise than by their gallantries, and potions, like abductions by masked men and imprisonments in subterranean dungeons, have even passed out of our romances. All these must be relegated to the Confessional of the Black Penitents or the ruins of the Castle of Udolpho.

Chapter XIX: The Mastery of the Sun

WE come now to that number which is attributed in the Tarot to the sign of the sun. The denary of Pythagoras and the triad multiplied by itself represent wisdom in its application to the absolute. It is with the absolute, therefore, that we are concerned here. To discover the absolute in the infinite, the indefinite, and the finite, such is the great work of the sages, that which is termed by Hermes the work of the sun. To find the immovable foundations of true religious faith, of philosophical truth, and of metallic transmutation, this is the whole secret of Hermes, this is the philosophical stone. Now, this stone is both one and manifold; it is decomposed by analysis and recomposed by synthesis. In the analysis it is a powder, the alchemical powder of projection; before the analysis and in the synthesis, it is a stone. The philosophical stone, say the masters, must not be exposed to the air, nor to the eyes of the profane; it must be kept in concealment and preserved carefully in the most secret receptacle of the laboratory, the key of the place being always carried upon the person.

He who possesses the great arcanum is truly king and is above any king, for he is inaccessible to all fears and to all vain hopes. In any malady of soul and body, a single fragment broken from the precious stone, a single grain of the divine powder, are more than sufficient for their cure. "He that hath ears to hear, let him hear," as the Master said.

Salt, sulphur and mercuries are only accessory elements and passive instruments of the great enterprise. Everything depends, as we have said, upon the interior *magnes* of Paracelsus. The work consists entirely in *projection*, and projection is accomplished perfectly by the effective and realizable intelligence of a single word. There is but one important operation, and that is *sublimation*, which is nothing else, according to Geber, but the elevation of the dry substance by means of fire, with adherence to its proper vessel. He who is desirous of understanding the great word and of possessing the great arcanum, after studying the principles of our Doctrine, should read the Hermetic philosophers carefully, and he will doubtless attain initiation, as others have attained it; but for the key of their allegories he must take the one dogma of Hermes, contained in the Emerald Table, and to classify the knowledge and direct operation he must follow the order indicated in the kabbalistic alphabet of the Tarot, of which an absolute and complete explanation will be given in the last chapter of this work.

Among the rare and priceless treatises which contain the mysteries of the great arcanum, the "Chemical Pathway or Manual" of Paracelsus must be placed in the first rank, as comprising all the mysteries of demonstrative physics and the most secret kabbalah. This unique manuscript is preserved in the Vatican Library; a copy was transcribed by Sendivogius and was used by Baron Tschoudy when composing the Hermetic Catechism contained in his work entitled "The Blazing Star." This catechism, which we point out to instructed kabbalists as a substitute for the incomparable treatise of Paracelsus, expounds all the essential principles of the great work in a form so clear and complete that a person must be absolutely wanting in the quality of occult understanding if he fail in attaining the absolute truth by its study. We shall now give a succinct analysis of this work, together with a few words by way of commentary.

Raymond Lully, one of the grand and sublime masters of science, says that before we can make gold we must have gold. Out of nothing we can make nothing; wealth is not created absolutely: it is increased and multiplied. Hence, let aspirants to knowledge understand thoroughly that neither miracles nor jugglers' feats are required of the adept. Hermetic science, like all real sciences, is mathematically demonstrable. Even its material results are as exact as a well-worked equation. Hermetic gold is not only a true doctrine, a shadowless light, truth unalloyed with falsehood; it is also material, actual, pure gold, the most precious which can be found in the veins of the earth; but the living gold, living sulphur, or true fire of the philosopher, must be sought in the house of mercury. This fire feeds on air; to express its attractive and expansive power, a better comparison is impossible than that of lightning, which primally is a dry and terrestrial exhalation united to humid vapor, and afterwards, assuming an igneous nature, in virtue of its exaltation, acts on its inherent humidity, which it attracts and transmutes into its own nature, after which it falls rapidly to earth, where it is drawn by a fixed nature similar to its own. These words, enigmatic in form but clear in essence, express openly what the philosophers understand by their mercury fructified by sulphur, becoming the master and regenerator of salt. It is AZOTH, universal *magnesia*, the great magical agent, the astral light, the light of life, fertilized by animic force, by intellectual energy, which they compare to sulphur on account of its affinities with divine fire. As to salt, it is absolute matter. All that is material contains salt, and all salt can be converted into pure gold by the combined action of sulphur and mercury, which at times act with such a swiftness that transmutation can take place in an instant, or in an hour, without labor for the operator and almost without expense; at other times, when the tendencies of the atmospheric media are more contrary, the operation requires several days, months, and occasionally, even years.

As we have already said, there are two palmary natural laws – two essential laws – which, balanced one against another, produce the universal equilibrium of things. These are fixity and motion, analogous to truth and discovery in philosophy, and in absolute conception to necessity and liberty, which are the very essence of God. The Hermetic philosophers give the name of fixed to all that is ponderable, to all that tends by its nature to central rest and immobility; whatsoever obeys more naturally and readily the law of motion, they term volatile; and they compose their Stone by analysis, that is, the volatilization of the fixed; then by synthesis, that is, the fixation of the volatile, which they operate by applying to the fixed, called their salt, sulphurated mercury or light of life, directed and rendered omnipotent by a secret operation. They possess themselves in this manner of all nature, and their stone is found wherever there is salt, which is the equivalent to saying that no substance is foreign to the great work, and that even the most apparently contemptible and vile matters can be changed into gold, which is true in this sense, as we have said, that all contain the fundamental salt, represented in our emblems by the cubic stone itself, as may be seen in the symbolic and universal frontispiece to the keys of Basil Valentine. To know how to extract from all matter the pure salt which is concealed in it is to possess the secret of the stone. It is therefore, a saline stone, which the *od*, or universal astral light, decomposes or recomposes. It is one and many, for, like ordinary salt, it can be dissolved and incorporated with other substances. Obtained by analysis, it may be termed the *universal sublimate*; recovered by the synthetic way, it is the veritable *panacea* of the ancients, for it cures all diseases, whether of soul or body, and is termed in an eminent manner, the medicine of all nature. When, by means of absolute initiation, we can dispose of the forces of the universal agent, this stone is always to our hand, for its extraction is then a simple and easy operation, far different from projection or metallic realization. The stone in its sublimated state must not be exposed to the air, which might dissolve it and spoil its virtue. Moreover, to inhale its exhalations is not devoid of danger. The wise man more readily conserves it in the natural envelopes, knowing that he can extract it by a single effort of his will, and a single application of the universal agent to the envelopes, which kabbalists term shells. To express hieroglyphically this law of prudence, the sages of Egypt ascribed to their mercury, personified in as Hermanubis, a dog's head, and to their Sulphur, represented by the Baphomet of the temple or prince of the Sabbath, that goat's head which brought such odium upon the occult associations of the middle ages.

For the mineral work, the first matter is exclusively mineral, but it is not a metal. It is a metallized salt. This matter is called vegetable, because it resembles a fruit, and animal, because it produces a kind of milk and blood. It alone contains the fire by which it must be dissolved.

Chapter XX: The Thaumaturge

WE have defined miracles as the natural effects of exceptional causes. The immediate action of the human will upon the body, or at least that action exercised without visible means, constitutes a miracle in the physical order. The influence exercised upon wills or intelligences, either suddenly or within a given time, and capable of subjugating thoughts, changing the most determined resolutions, paralyzing the most violent passions – this influence constitutes a miracle in the moral order. The common error concerning miracles is to regard them as effects without causes, contradictions of nature, sudden vagaries of the divine mind, not seeing that a single miracle of this class would destroy the universal harmony, and reduce the universe to chaos. There are miracles which are impossible, even for God, namely, those that involve absurdity. Could God be absurd for one instant, neither Himself nor the world would be in existence the moment following. To expect from the divine arbiter an effect having a disproportionate cause, or even no cause at all, is what is called tempting God; it is casting one's self into the void. God operates by His works – in heaven by angels and on earth by men. Hence, in the circle of angelic action, the angels can perform all that is possible for God, and in the human circle of action men can dispose equally of divine omnipotence. In the heaven of human conceptions, it is humanity which creates God, and men think that God has made them in His image because they have made Him in theirs. The domain of man is all corporeal and visible nature on earth, and if he cannot rule suns and stars, he can at least calculate their motion, compute their distances, and identify his will with their influence; he can modify the atmosphere, act up to a certain point upon the seasons, heal or harm his neighbors, preserve life and inflict death, the conservation of life, including resurrection in certain cases, as already established. The absolute in reason and volition is the greatest power which can be given any man to attain, and it is by means of this power that he performs what astonishes the multitude under the name of miracles.

The most perfect purity of intention is indispensable to the thaumaturge, and in the next place a favorable current and unlimited confidence. The man who has come to fear nothing and desire nothing is master of all. This is the meaning of that beautiful allegory of the Gospel, wherein the Son of God, thrice victor over the unclean spirit, is ministered to by angels in the wilderness. Nothing on earth withstands a free and rational will. When the wise man says, "I will," it is God Himself who wills, and all that He commands takes place. It is the knowledge of the physician, and the confidence placed in him, which constitute the virtue of his prescriptions, and thaumaturgy is the only real and efficacious remedy. Hence occult therapeutics are apart from all vulgar medication. It chiefly makes use of words and insufflations, and communicates by will a various virtue to the simplest substances – water, oil, wine camphor, salt. The water of homœopathists is truly a magnetized and enchanted water, which works by means of faith. The dynamic substances added in, so to speak, infinitesimal quantities are consecrations and signs of the physician's will.

What is vulgarly called charlatanism is a great means of real success in medicine, assuming that it is sufficiently skillful to inspire great confidence and to form a circle of faith. In medicine, above all, it is faith which saves. There is scarcely a village which does not possess its male or female compounder of occult medicine, and these people are almost everywhere, and invariably, more successful incomparably than physicians approved by the faculty. The remedies they prescribe are often strange or ridiculous, and hence answer all the better, for they exact and realize more faith on the part of patients and operators. An old merchant of our acquaintance, a man of eccentric character and exalted religious sentiment, after retiring from business, set himself to exercise gratuitously, and out of Christian charity, occult medicine in one of the Departments of France. His sole specifics were oil, insufflations and prayers. The institution of a lawsuit against him for the illegal exercise of medicine established in public knowledge that ten thousand cures had been attributed to him in the space of about five years, and that the number of his believers increased in proportions calculated to alarm all the doctors of the district. We saw also at Mans a poor nun who was regarded as slightly demented, but she healed, nevertheless, all diseases in the surrounding country by means of an elixir and plaster of her own invention. The elixir was taken internally, the plaster was applied outwardly, so that nothing escaped this universal panacea. The plaster never stuck upon the skin save at the place where its application was necessary, and it rolled up and fell off by itself – such at least was asserted by the good sister and declared to be the case by the sufferers. This thaumaturge was also subjected to prosecution, for she impoverished the practice of all the doctors round about her; she was rigidly cloistered, but it was soon found necessary to produce her at least once a week, and on the day for her consultations we have seen Sister Jane-Frances surrounded by the country folk, who had arrived overnight, awaiting their turn, lying at the convent gate; they had slept upon the ground, and tarried only to receive the elixir and plaster of the devoted sister. The remedy being the same in all diseases, it would appear needless for her to be acquainted with the cases of her patients, but she listened to them invariably with great attention, and only dispensed her specific after learning the nature of the complaint. There was the magical secret. The direction of the intention imparted its special virtue to the remedy, which was insignificant in itself. The elixir was spiced brandy mixed with the juice of bitter herbs; the plaster was a compound analogous to theriac as regards color and smell; it was possibly electuary Burgogne pitch, but whatever the substance, it worked wonders, and the wrath of the rural folk would have been visited on those who questioned the miracles of their nun. Near Paris, also, we knew of an old gardener thaumaturge who accomplished marvelous cures by putting in his phials the juice of all the herbs of St. John. He had, however, a skeptical brother, who derided the sorcerer, and the poor gardener, overwhelmed by the sarcasms of this infidel, began to doubt himself, whereupon all the miracles ceased, the sufferers lost confidence, and the thaumaturge, slandered and despairing, died mad. The Abbé Thiers, curé of Vibraie, in his curious "Treatise concerning Superstitions," records that a woman, afflicted with an apparently aggravated ophthalmia, having been suddenly and mysteriously cured, confessed to a priest that she had betaken herself to magic. She had long importuned a clerk, whom she regarded as a magician, to give her a talisman that she might wear, and he had at length delivered her a scroll of parchment, advising her at the same time to wash three times daily in fresh water. The priest made her give up the parchment, on which were these words: *Eruat diabolus oculos tuos et repleat stercoribus loca vacantia.* He translated them to the good woman, who was stupefied; but, all the same, she was cured.

Insufflation is one of the most important practices of occult medicine, because it is a perfect sign of the transmission of life. To inspire, as a fact, means to breathe upon some person or thing, and we know already, by the one doctrine of Hermes, that the virtue of things has created words, and that there is an exact proportion between ideas and speech, which is the first form and verbal realization of ideas. The breath attracts or repels accordingly as it is warm or cold. The warm breathing corresponds to positive electricity, and the cold breathing to negative electricity. Electrical and nervous animals fear the cold breathing, and the experiment may be made upon a cat, whose familiarities are importunate. By fixedly regarding a lion or tiger and blowing in their face, they would be so stupefied as to be forced to retreat before us. Warm and prolonged insufflation restores the circulation of the blood, cures rheumatic and gouty pains, re-establishes the balance of the humours, and dispels lassitude. When the operator is sympathetic and good, it acts as a universal sedative. Cold insufflation soothes pains occasioned by congestions and fluidic accumulations. The two breathings must, therefore, be used alternately, observing the polarity of the human organism, and acting in a contrary manner upon the poles, which must be subjected successfully to an opposite magnetism. Thus, to cure an inflamed eye, the one which is not affected must be subjected to a warm and gentle insufflation, cold insufflation being practiced upon the suffering member at the same distance and in the same proportion. Magnetic passes have a similar effect to insufflations, and are a real breathing by transpiration and radiation of the interior air, which is phosphorescent with vital light; slow passes constitute a warm breathing which fortifies and raises the spirits; swift passes are a cold breathing of dispersive nature, neutralizing tendencies to congestion. The warm insufflation should be performed transversely, or from below upward; the cold insufflation is more effective when directed downward from above.

We breathe not only by means of mouth and nostrils; the universal porousness of our body is a true respiratory apparatus, inadequate undoubtedly, but most useful to life and health. The extremities of the fingers, where all the nerves terminate, diffuse or attract the astral light accordingly as we will. Magnetic passes without contact are a simple and slight insufflation; contact adds sympathetic and equilibrating impression; it is good and even necessary, to prevent hallucinations at the early stages of somnambulism, for it is a communion of physical reality which admonishes the brain and recalls wandering imagination; it must not, however, be too prolonged when the object is merely to magnetize. Absolute and prolonged contact is useful when the design is incubation or massage rather than magnetism properly so called. We have given some examples of incubation from the most revered book of the Christians; they all refer to the cure of apparently incurable lethargies, as we are induced to term resurrections. Massage is still resorted to in the east, where it is practiced with great success at the public baths. It is entirely a system of frictions, tractions, and pressures, practiced slowly along the whole length of members and muscles, the result being renewed equilibrium in the forces, a feeling of complete repose and well-being, with a sensible restoration of activity and vigor.

The whole power of the occult physician is in the conscience of his will, while the whole art consists in exciting the faith of his patient. "If you have faith," said the Master, "all things are possible to him who believes." The subject must be dominated by expression, tone, gesture; confidence must be inspired by a fatherly manner, and cheerfulness stimulated by seasonable and sprightly conversations. Rabelais, who was a greater magician than he seemed, made pantagruelism his special panacea. He compelled his patients to laugh, and all the remedies he subsequently gave them succeeded the better in consequence; he established a magnetic sympathy between himself and them, by means of which he communicated to them his own confidence and good humor; he flattered them in his prefaces, termed them his precious, most illustrious patients, and dedicated his books to them. So are we convinced the Gargantua and Pantagruel cured more black humours, more tendencies to madness, more atrabilious whims, at that epoch of religious animosities and civil wars, than the whole Faculty of medicine could boast. Occult medicine is essentially sympathetic. Reciprocal affection, or at least real good will, must exist between doctor and patient. Syrups and juleps have very little inherent virtue; they are what they become through the mutual opinion of operator and subject; hence homœopathic medicine dispenses with them and no serious inconvenience follows. Oil and wine, combined with salt or camphor, are sufficient for the healing of all afflictions, and for all external frictions or soothing applications, oil and wine are the chief medicaments of Gospel tradition. They formed the balm of the Good Samaritan, and in the Apocalypse, when describing the last plagues, the prophet prays the avenging powers to spare these substances, that is, to leave a hope and a remedy for so many wounds. What we term extreme unction was the pure and simple practice of the Master's traditional medicine, both for the early Christians and in the mind of the apostle Saint James, who has included the precept in his epistle to the faithful of the whole world. "If any man sick among you," he writes, "let him call in the priests of the church, and let them pray over him, anointing him with oil in the name of the Lord." This divine therapeutic science was lost gradually, and Extreme Unction came to be regarded as a religious formality necessary as a preparation for death. At the same time, the thaumaturgic virtue of consecrated oil could not be altogether effaced from remembrance by the traditional doctrine, and it is perpetuated in the passage of the catechism which refers to Extreme Unction. Faith and charity were the most signal healing powers among the early Christians. The source of most diseases is in moral disorders; we must begin by healing the soul, and then the cure of the body will follow quickly.

Chapter XXI: The Science of Prophets

THIS chapter is consecrated to divination, which, in its broadest sense, and following the grammatical significance of the word, is the exercise of divine power, and the realization of divine knowledge. It is the priesthood of the magus. But divination, in general opinion, is concerned more closely with the knowledge of hidden things. To know the most secret thoughts of men; to penetrate the mysteries of past and future; to evoke age by age the exact revelation of effects by the precise knowledge of causes; this is what is universally called divination. Now, of all mysteries of nature, the most profound is the heart of man; but at the same time nature forbids its depth to be inaccessible. In spite of the deepest dissimulation, despite the most skillful policy, she herself traces, and makes plain in the bodily form, in the light of glances, in movements, in carriage, in voice, a thousand tell-tale indices. The perfect initiate has no need of these indices; he perceives the truth in the light; he senses an impression which makes known the whole man; his glance penetrates hearts, he may even feign ignorance to disarm the fear or hatred of the wicked whom he knows too well. A man of bad conscience thinks always that he is being accused or suspected; he recognizes himself in a touch of collective satire, he applies it in that whole satire to himself, and cries loudly that he is calumniated. Ever suspicious, but as curious as he is apprehensive, in the presence of the magus he is like Satan of the parable, or like those scribes who questioned tempting. Ever stubborn and ever feeble, what he fears above all is the recognition that he is in the wrong. The past disquiets him, the future alarms him; he seeks to compound with himself and to believe himself a well-placed and virtuous man. His life is a perpetual struggle between good aspirations and evil habits; he thinks himself a philosopher after the manner of Aristippe or Horace in accepting all the corruption of his time as a necessity which he must undergo; he distracts himself with some philosophical pastime, and appropriates the protecting smile of Mecænas to persuade himself that he is not simply a batterer on famine like Verres or a parasite of Trimalcion. Such men are always mercenaries, even in their good works. They decide to make a gift to a public charity, and they postpone it to get the interest. The type which I am describing is not an individual but a class of men with which the magus is liable to come frequently in contact, especially in our own century. Let him follow their own example by mistrusting them, for they will be invariably his most compromising friends and most dangerous enemies.

The public exercise of divination is unbecoming at the present period in a veritable adept, for he would be frequently driven to jugglery and feats of skill in order to preserve his clients and astonish his public. Accredited diviners, both male and female, have always secret spies, who instruct them as to the private life or habits of those who consult them. A code of signals is established between cabinet and antechamber; an unknown client at his first visit receives a number; a day is arranged, and he is followed; doorkeepers, neighbors, servants are engaged in gossip, and details are thus arrived at which overwhelm simple minds, leading them to invest an impostor with the reverence which should be reserved for true science and genuine divination.

The divination of events to come is possible only in the case of those the realization of which is in some sense contained in their cause. The soul, scrutinizing by means of the whole nervous system the circle of the astral light which influences a man and from him receives an influence, the soul of the diviner, we repeat, can compass by a single intuition all the loves and hatreds which that man has evoked about him; it can read his intentions in his thought, foresee obstacles that he will encounter, possibly the violent death which awaits him; but it cannot foresee his private, voluntary, capricious determinations of the moment following the consultation, unless, indeed, the ruse of the diviner itself prepares the fulfilment of the prophecy. For example, you say to a woman who is becoming *passé*, and is anxious to secure a husband: You will be present this evening or tomorrow evening at such or such a performance, and you will there see a man who will be to your liking. This man will observe you, and by a curious combination of circumstances the result will be a marriage.

You may count on the lady going, you may count on her seeing a man and believing that he has noticed her, you may count on her anticipating marriage. It may not come to that in the end, but she will not lay the blame on you, because she would be giving up the opportunity for another illusion; on the contrary, she will return perseveringly to consult you.

We have said that the astral light is the great book of divinations; the faculty of reading therein is either natural or acquired, and there are hence two classes of seers, the instinctive and the initiated. For this reason, children, uneducated people, shepherds, even idiots, have more aptitude for natural divination than scholars and thinkers. The simple herd-boy, David, was a prophet even as Solomon, king of kabbalists and magi. The perceptions of instinct are often as certain as those of science; the persons least clairvoyant in the astral light are those who reason most. Somnambulism is a state of pure instinct, and hence somnambulists require to be directed by a seer of science; sceptics and reasoners only lead them astray. Divinatory vision operates only in the ecstatic state, to arrive at which state, doubt and illusion must become impossible by enchaining or putting to sleep thought. The instruments of divination are hence only methods of magnetizing ourselves and of self-diversion from exterior light, so that we may pay attention to the interior light alone. It was for this reason that Apollonius completely enveloped himself in a woolen mantle, and fixed his eyes on his navel in the gloom. The magical mirror of Dupotet is kindred to the device of Apollonius. Hydromancy and vision in the thumb-nail, when it has been polished and blackened, are varieties of the magical mirror. Perfumes and evocations stupefy thought; water and the color black absorb the visual rays; a kind of dazzlement and vertigo ensue, followed by lucidity in subjects who have a natural aptitude or are suitably disposed thereto. Geomancy and cartomancy are other means to the same end; combinations of symbols and numbers, which are at once fortuitous and necessary, bear enough resemblance to the chances of destiny for the imagination to perceive realities by the pretext of such emblems. The more the interest is excited, the greater is the desire to see; the fuller the confidence in the intuition, the more clear the vision becomes. To combine the points of geomancy on chance or to set out the cards for trifling is to jest like children: the lots become oracles only when they are magnetized by intelligence and directed by faith.

Of all oracles, the Tarot is the most astounding in its answers, because all possible combinations of this universal key of the kabbalah give oracles of science and of truth as their solutions. The Tarot was the sole book of the ancient magi; it is the primitive Bible, as we shall prove in the following chapter, and the ancients consulted it as the first Christians at a later date consulted the Sacred Lots, that is, Bible verses selected by chance and determined by thinking of a number. Mlle. Lenormand, the most celebrated of our modern fortune-tellers, was unacquainted with the science of the Tarot, or knew it only by derivation from Etteilla, whose explanations are shadows cast upon a background of light. She knew neither high magic nor the kabbalah, but her head was filled with ill-digested erudition, and she was intuitive by an instinct, which deceived her rarely. The works she left behind her are Legitimist tomfoolery, ornamented with classical quotations, but her oracles inspired by the presence and magnetism of those who consulted her, were often astounding. She was a woman in whom extravagance of imagination and mental rambling were substituted for the natural affections of her sex; she lived and died a virgin, like the ancient druidesses of the isle of Sayne. Had Nature endowed her with beauty, she might have played easily at a remoter epoch the part of a Melusine or a Velléda.

The more ceremonies employed in the practice of divination, the more we stimulate imagination both in ourselves and in those who consult us. The Conjuration of the Four, the Prayer of Solomon, the magic sword to disperse phantoms, may then be resorted to with success; we should also evoke the genius of the day and hour of operation, and offer him a special perfume; next we should enter into magnetic and intuitive correspondence with the consulting person, inquiring with what animal he is in sympathy and with what in antipathy, as also concerning his favorite flower or color. Flowers, colors and animals connect in analogical classification with the seven genii of the kabbalah. Those who love blue are idealists and dreamers; lovers of red are material and passionate; those who love yellow are fantastic and capricious; lovers of green are frequently commercial and crafty; the friends of black are influenced by Saturn; the rose is the color of Venus, etc. Lovers of the horse are hard-working, noble in character, and at the same time yielding and gentle; friends of the dog are affectionate and faithful; those of the cat are independent and libertine. Frank persons hold spiders in special horror; those of haughty nature are antipathetic to the serpent; upright and fastidious persons cannot tolerate rats and mice; the voluptuous loathe the toad, because it is cold, solitary, hideous and miserable. Flowers have analogous sympathies to those of animals and colors, and as magic is the science of universal analogies, a single taste, one tendency, in a given person, enables all the rest to be divined: it is an application of the analogical anatomy of Cuvier to phenomena in the moral order.

The physiognomy of face and body, the wrinkles on the brow, the lines on hands, furnish the magus with precious indications. Metoposcopy and chiromancy have become separate sciences; their findings, purely empirical and conjectural, have been compared, examined and then united into a body of doctrine by Goclenius, Belot, Romphile, Indagine, and Taisnier. The work of the last-mentioned writer is the most important and complete; he combines and criticizes the observations and conjectures of all the others. A modern investigator, the Chevalier D'Arpentigny, has imparted to chiromancy a fresh degree of certitude by his remarks on the analogies which really exist between the characters of persons and the form of their hands as a whole or in detail. This new science has been developed and verified by an artist who is also a man of letters, rich in originality and skill.

The disciple has surpassed the master, and our amiable and spiritual Desbarrolles, one of those travelers with whom our great novelist Alexandre Dumas delights to surround himself in his cosmopolitan romances, is quoted already as a veritable magician in chiromancy.

The consulting person should also be questioned upon his habitual dreams; dreams are the reflection of life, both interior and exterior. The old philosophers paid them great attention; the patriarchs regarded them as certain revelations; most religious revelations have been given in dreams. The monsters of perdition are nightmares of Christianity, and as the author of Smarra has ingeniously remarked, never could pencil or chisel have produced such prodigies if they had not been beheld in sleep. We should beware of persons whose imagination continually reflects deformities. Temperament is, in like manner, manifested by dreams, and as this exercises a permanent influence upon life, it is necessary to be well acquainted therewith if we would conjecture a destiny with certitude. Dreams of blood, of enjoyment, and of light indicate a sanguine temperament; those of water, mud, rain, tears, are occasioned by a more phlegmatic disposition; fire by night, darkness, terrors, specters, belong to the bilious and melancholic. Synesius, one of the greatest Christian bishops of the first centuries, the disciple of that beautiful and pure Hypatia who was massacred by fanatics after presiding gloriously over the school of Alexandria, in the inheritance of which school Christianity should have shared – Synesius, lyric poet like Pindar and Callimachus, priest like Orpheus, Christian like Spiridion of Tremithonte – has left us a treatise on dreams which has been supplied with a commentary by Cardan. No one concerns themselves now with these magnificent researches of the mind because successive fanaticisms have well-nigh forced the world to despair of scientific and religious rationalism. St.

Paul burned Trismegistus; Omar burned the disciples of Trismegistus and St. Paul. O persecutors! O incendiaries! O scoffers! When will ye end your work of darkness and destruction?

One of the greatest magi of the Christian era, Trithemius, irreproachable abbot of a Benedictine monastery, learned theologian, and master of Cornelius Agrippa, has left among his unappreciated and inestimable works, a treatise entitled, *De septem secundeis, id est intelligentiis sive spiritibus orbes post Deum moventibus*. It is a key of all prophecies new or old, a mathematical, historical and simple method of surpassing Isaiah and Jeremiah in the prevision of all great events to come. The author in bold outline sketches the philosophy of history, and divides the existence of the entire world between seven genii of the kabbalah. It is the grandest and widest interpretation ever made of those seven angels of the Apocalypse who appear successively with trumpets and cups to pour out the word and its realization upon earth. The duration of each angelic reign is 354 years and 4 months, beginning with that of Orifiel, the angel of Saturn, on the 13th of March, for, according to Trithemius, this was the date of the world's creation; it was a period of savagery and darkness. Next came the reign of Anaël, the spirit of Venus, on the 24th of June, in the year of the world 354, when love began to be the instructor of mankind; it created the family, while the family led to association and the primitive city. The first civilizers, were poets inspired by love; presently the exaltation of poetry produced religion, fanaticism and debauchery, culminating subsequently in the deluge. This state of things continued till the 25th of October, being the eighth month of the year A.M. 708, when the reign of Zachariel, the angel of Jupiter, was inaugurated, under whose guidance men began to acquire knowledge and dispute the possession of lands and dwellings. It was also the epoch of the foundation of towns and the limitation of empires; its consequences were civilization and war. The need of commerce began, furthermore, to be felt, at which time – namely, the 24th of February, A.M. 1063 – was inaugurated the reign of Raphael, angel of Mercury, angel of science and of the word, of intelligence and industry. Then letters were invented, the first language being hieroglyphic and universal, a monument of which has been preserved in the book of Enoch, Cadmus, Thoth, and Palamedes; the kabbalistic clavicle adopted later on by Solomon, the mystical book of the Theraphim, Urim and Thummim, the primeval Genesis of the Zohar, and of William Postel, the mystical wheel of Ezekiel, the *rota* of the Kabbalists, the Tarot of Magi and Bohemians. Then were arts invented, and navigation was attempted for the first time; relations extended, wants multiplied and there followed speedily an epoch of general corruption, preceding the universal deluge, under the reign of Samaël, angel of Mars, which was inaugurated on the 26th of June, A.M. 1417. After long stupefaction, the world strove towards a new birth under Gabriel, the angel of the moon, whose reign began on the 28th of March, A.M. 1771, when the family of Noah became multiplied and re-peopled the whole earth, after the confusion of Babel, until the reign of Michael, angel of the sun, which commenced on the 24th of February, A.M. 2126, to which epoch must be referred the origin of the first dominations, the empire of the children of Nimrod, the birth of sciences and religions, and the first conflicts between despotism and liberty. Trithemius pursues this curious study throughout the ages, and at corresponding epochs exhibits the recurrence of ruins; then civilization, born anew by means of poetry and love; empires, reconstituted by the family, enlarged by commerce, destroyed by war, repaired by universal and progressive civilization, subsequently absorbed by great empires, which are the syntheses of history. The work of Trithemius, from this point of view, is more comprehensive and independent than that of Bossuet, and is a key absolute to the philosophy of history. His exact calculations lead him to the month of November in the year 1879, epoch of the reign of Michael and the foundation of a new universal kingdom, prepared by three centuries and a half of anguish, and a like period of hope, coinciding precisely with the sixteenth, seventeenth, eighteenth, and first part of the nineteenth centuries for the lunar twilight and expectation, with the fourteenth, thirteenth, twelfth, and second half of the eleventh centuries for the ordeals, the ignorance, the sufferings, and the scourges of all nature. We see, therefore, according to this calculation, that in 1879 – that is, in twenty-four years' time, a universal empire will be founded, and will secure peace to the world. This empire will be political and religious; it will offer a solution for all problems agitated in our own days, and will endure for 354 years and 4 months, after which it will be succeeded by the return of the reign of Orifiel, an epoch of silence and night. The coming universal empire, being under the reign of the sun, will belong to him who holds the keys of the East, which are now being disputed by the princes of the world's four quarters. But intelligence and activity are the forces which rule the sun in the superior kingdoms, and the nation which now possesses the initiative of intelligence and life will possess also the keys of the East, and will establish the universal kingdom. To do this it may previously have to undergo a cross and martyrdom analogous to those of the Man-God; but, dead or living, among nations its spirit will prevail, and all peoples will acknowledge and follow in four-and-twenty years the standard of France, ever victorious, or miraculously raised from the dead. Such is the prophecy of Trithemius, confirmed by all our previsions, and grounded in all our hopes.

Éliphas Lévi

Chapter XXII: The Book of Hermes

WE approach the end of our work, and must here give the universal key and utter the final word. The universal key of magical works is the key of all ancient religious dogmas, – the key of Kabbalah and the Bible, the little key of Solomon. Now, this clavicle, regarded as lost for centuries, has been recovered by us, and we have been able to open the sepulchers of the ancient world, to make the dead speak, to behold the monuments of the past in all their splendor, to understand the enigmas of every sphinx, and to penetrate all sanctuaries. Among the ancients the use of this key was permitted to none but the high priests, and even its secret was confided only to the flower of initiates. Now, this was the key in question: A hieroglyphic and numeral alphabet, expressed by characters and numbers, multiplied by four symbols, and connected with twelve figures representing the twelve signs of the zodiac, plus the four genii of the cardinal points.

The symbolical tetrad, represented in the mysteries of Memphis and Thebes by the four forms of the sphinx – the man, eagle, lion, and bull – corresponded with the four elements of the old world, water being signified by the cup held by the man or aquarius; air by the circle or numbus surrounding the head of the celestial eagle; fire by the wood which nourishes it, by the tree fructifying in the heat of earth and sun, and, finally, by the scepter of royalty, which the lion typifies; earth by the sword of Mithras, who each year immolates the sacred bull, and, together with its blood, pours forth that sap which gives increase to all fruits of earth. Now, these four signs, with all their analogies, explain the one word hidden in all sanctuaries, that word which the bacchanates seemed to divine in their intoxication when they worked themselves into frenzy for IO EVOHE. What, then, was the meaning of this mysterious term? It was the name of the four primitive letters of the mother tongue: the JOD, symbol of the vine, or paternal scepter of Noah; the HE, type of the cup of libations and also of maternity; the VAU, which joins the two, and was depicted in India by the great and mysterious lingam. Such was the triple sign of the triad in the divine word; then the mother letter appeared a second time to express the fecundity of nature and woman, and to formulate the doctrine of universal and progressive analogides descending from causes to effects, and ascending from effects to causes. Moreover, the sacred word was not pronounced; it was spelt, and read off in four words, which are the four sacred words – JOD HE VAU HE.

The learned Gaffarel regards the *teraphim* of the Hebrews, by means of they consulted the oracles of the *urim* and *thummim*, as the figures of the four kabbalistic animals, which symbols, as we shall presently show, were summed up in the sphinxes or cherubs of the ark. In connection with the usurped Teraphim of Michas, he cites a curious passage from Philo, which is a complete revelation as to the ancient and sacerdotal origin of our TAROTS. Gaffarel thus expresses himself: "He (Philo the Jew), speaking of the history concealed in the before-mentioned chapter of Judges, says that Michas made three images of young boys and three young calves, three also of a lion, an eagle, a dragon, a dove, all of fine gold and silver; so that if any one sought him to discover a secret concerning his wife, he interrogated the dove;

concerning his children, the young boy; concerning wealth, the eagle; concerning strength and power, the lion; concerning fecundity, the cherub or bull; concerning length of days, the dragon." This revelation of Philo, though deprecated by Gaffarel, is for us of the highest importance. Here, in fact, is our key of the tetrad, and here also the images of the four symbolical animals found in the twenty-first key of the Tarot; that is, at the third septenary, thus repeating and summarizing all the symbolism expressed by the three septenaries superposed; next, the antagonism of colors expressed by the dove and the dragon; the circle or ROTA, formed by the dragon or serpent to typify length of days; finally, the kabbalistic divination of the entire Tarot, as practiced in later days by the Egyptian Bohemians, whose secrets were divined and recovered imperfectly by Etteilla.

We see in the Bible that the high priests consulted the Lord on the golden table of the holy ark, between the cherubs, or bull-headed and eagle-winged sphinx; that they consulted by the help of the Theraphim, Urim, and Thummim, and by the Ephod. Now, it is known that the Ephod magical square of twelve numbers and twelve words engraved on precious stones. The word *Teraphim* in Hebrew signifies hieroglyphs or figured signs; the Urim and Thummim were the above and beneath, the east and the west, the yes and no, and these signs corresponded to the two pillars of the Temple, JAKIN and BOHAS. When, therefore, the high priest wished to consult the oracle, he drew by lot the Theraphim or tablets of gold, which bore the images of the four sacred words, and placed them by threes round the rational or Ephod; that is, between the two onyx stones which served as clasps to the little chains of the Ephod. The right onyx signified Gedulah, or mercy and magnificence; the left referred to Geburah, and signified justice and anger. If, for example, the sign of the lion were found on the left side of the stone which bore the name of the tribe of Judah, the high priest would read the oracle thus: "The staff of the Lord is angered against Judah." If the Theraphim represented the man or cup, and were also found on the left, near the stone of Benjamin, the high priest would read: "The mercy of the Lord is weary of the offences of Benjamin, which violate Him in His love. Whence he will pour out on him the chalice of his wrath," etc. When the sovereign priesthood ceased in Israel, when all oracles were silenced in the presence of the Word made man, and speaking by the mouth of the most popular and mildest of sages, when the ark was lost, the sanctuary profaned, and the temple destroyed, the mysteries of the Ephod and Theraphim, no longer traced on gold and precious stones, were written, or, rather, drawn, by some learned kabbalists on ivory, parchment, gilt and silvered copper, and, finally, on simple cards, which were always suspected by the official Church as enclosing a dangerous key to its mysteries. Hence came those Tarots, the antiquity of which, revealed to the erudite Court de Gebelin by the science of hieroglyphs and numbers, so exercised later the doubtful perspicacity and persistent investigation of Etteilla. Court de Gebelin, in the eighth volume of his "Primeval World," gives the figure of the twenty-two keys and four aces of the Tarot, and demonstrates their perfect analogy with all symbols of the highest antiquity. He subsequently endeavors to supply their explanation, and goes astray naturally, because he does not start from the universal and sacred tetragram, the IO EVOHE of the Bacchanalia, the JOD HE VAU HE of the sanctuary, the יהוה of the Kabbalah.

Etteilla or Alliette, preoccupied entirely by his system of divination and the material profit to be derived from it, Alliette, formerly barber, having never learned French, or even orthography, pretended to reform and thus appropriate the Book of THOT. In the Tarot, now become very scarce, which he engraved, we find the following naïve advertisement on the twenty eighth card – the eight of clubs: "Etteilla, professor of algebra and correctors (*sic*) of the *modern* blunders of the ancient book of Thot, lives in the Rue de l'Oseille, No. 48, Paris." Etteilla would have certainly done better not to have corrected the *blunders* of which he speaks; his books have degraded the ancient work discovered by Court de Gabelin into the domain of vulgar magic and fortune-telling by cards. He proves nothing who tries to prove too much; Etteilla furnishes another example of this old logical axiom; at the same time, his endeavors led him to a certain acquaintance with the Kabbalah, as may be seen in some rare passages of his unreadable works. The true initiates who were Etteilla's contemporaries, the Rosicrucians, for example, and the Martinists, who were in possession of the true Tarot, as a work of St. Martin proves, where the divisions are those of the Tarot, and this passage of an enemy of the Rosicrucians: "They pretend to the possession of a volume from which they can learn anything that can possibly be found in other books which now exist or may at any time be produced. This volume is their reason, in which they find the prototype of everything that exists by the facility of analyzing, making abstractions, forming a species of intellectual world, and creating all possible beings. See the philosophical, theosophical, microcosmic cards." (*Conspiracy against the Catholic Religion and Sovereigns*, by the author of *The Veil raised for the Curious*. Paris: Crapard., 1792). The true initiates, we repeat, who held the Tarot secret among their greatest mysteries, carefully refrained from protesting against the errors of Etteilla, and left him to revail instead of revealing the arcana of the true clavicles of Solomon. Hence it is not without profound astonishment that we have discovered intact and still unknown this key of all doctrines and all philosophies of the old world. I speak of it as a key, and such it truly is, having the circle of four decades as its ring, the scale of 22 characters as its trunk or body, and the three degrees of the triad as its wards; as such it was represented by Postel in his "Key of Things Kept Secret from the Foundation of the World." He indicates after the following manner the occult name of this key, which is known only to initiates:–

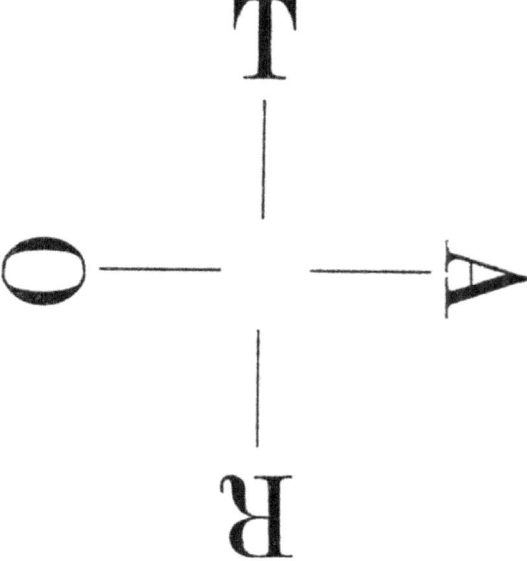

The word may read ROTA, thus signifying the wheel of Ezekiel, or TAROT, and then it is synonymous with the AZOTH of the Hermetic philosophers. It is a word which kabbalistically expresses the dogmatic and natural absolute; it is formed of the characters of the monogram of Christ, according to the Greeks and Hebrews. The Latin R or Greek P is found between the alpha and omega of the Apocalypse; the sacred Tau, image of the cross, encloses the who0le word, as previously represented in our Ritual. Without the Tarot, the magic of the ancients is a closed book, and it is impossible to penetrate any of the great mysteries of the Kabbalah. The Tarot alone interprets the magic squares of Agrippa and Paracelsus, as we may satisfy ourselves by forming these same squares with the keys of the Tarot, and reading off the hieroglyphs thus collected. These are seven magical squares of the planetary genii according to Paracelsus: –

Éliphas Lévi

SATURN.

2	9	4
7	5	3
6	1	8

JUPITER.

6	12	12	10
5	10	11	11
9	6	7	12
14	6	4	1

MARS.

14	10	22	22	18
20	12	7	20	2
8	17	9	9	8
12	3	9	5	26
11	23	8	6	11

THE SUN.

9	22	1	32	25	19
7	11	27	18	8	3
19	14	16	15	23	24
18	20	22	21	17	13
22	29	10	19	26	12
36	5	35	6	12	13

VENUS.

22	47	18	41	0	35	8
25	23	47	17	42	11	29
10	6	14	9	18	36	12
3	31	16	25	43	19	37
38	14	32	31	26	44	20
21	39	8	33	22	27	45
46	15	40	19	24	03	27

MERCURY.

8	52	39	5	24	61	66	11
49	15	14	52	52	12	10	56
41	43	22	14	45	19	18	48
33	34	35	29	20	38	39	25
40	6	27	59	31	30	31	33
17	47	55	28	25	43	42	24
9	51	53	12	13	51	00	16
64	12	15	61	61	6	7	47

THE MOON.

37	70	29	70	21	62	12	14	41
16	28	70	30	71	12	53	14	46
47	20	11	7	31	72	22	35	15
16	48	68	40	81	32	62	25	56
57	17	49	29	7	66	33	65	25
26	58	40	56	31	42	74	34	66
53	27	59	10	51	2	41	75	35
36	68	19	60	11	65	43	44	76
77	28	20	69	61	12	25	60	5

By adding each of the columns of these squares, you will obtain invariably the characteristic number of the planet, and, finding the explanation of this number by the hieroglyphs of the Tarot, you proceed to seek the sense of all the figures, whether triangular, square, or cruciform, that you find to be formed by the numbers. The result of this operation will be a complete and profound acquaintance with all allegories and mysteries concealed by the ancients under the symbol of each planet, or rather of each personification of influences, celestial or human, upon all events of life.

We have said that the twenty-two keys of the Tarot are the twenty-two letters of the primitive kabbalistic alphabet, and here follows a table of the variants of this alphabet according to divers Hebrew kabbalists:

א *Being, mind, man, or God; the comprehensible object; unity, mother of numbers, the first substance.*

All these ideas are hieroglyphically expressed by the figure of the JUGGLER. His body and arms constitute the letter aleph; round his head there is a nimbus in the form of ∞, the emblem of life and the universal spirit; in front of him are swords, cups, and pantacles; he uplifts the miraculous rod towards heaven. He has a youthful figure and curly hair, like Apollo or Mercury; the smile of confidence is on his lips and the look of intelligence in his eyes.

ב *The house of God and man, the sanctuary, the law, the Gnosis, Kabbalah, the occult church, the duad, wife, mother.*

Hieroglyph of the Tarot, THE FEMALE POPE; a woman crowned with a tiara, wearing the horns of the Moon and Isis, her head enveloped in a mantle, the solar cross on her breast, and holding a book on her knees, which she conceals with her mantle. A protestant author of a pretended history of Pope Joan has met with, and used, for good or bad, in the interests of his thesis, two curious and ancient figures of the female pope or sovereign priestess of the Tarot. These figures ascribe to her all the attributes of Isis; in one she is carrying and caressing her son Horus; in the other, she has long and thin hair; she is seated between the two pillars of the duad, has a sun with four rays on her breast, places one hand upon a book, and makes the sign of sacerdotal esotericism with the other – that is to say, she uplifts three fingers only, the two others being folded, to signify mystery; a veil is behind her head, and on each side of her chair the flowers of the lotus bloom upon the sea. I strongly commiserate the unlucky scholar who has seen in this antique symbol nothing but a monumental portrait of his pretended Pope Joan.

ג *The word, the triad, plenitude, fecundity, nature, generation in the three worlds.*

Symbol, THE EMPRESS, a woman, winged, crowned, seated, and uplifting a scepter with the orb of the world at its end; her sign is an eagle, image of the soul and of life. This woman is the Venus-Urania of the Greeks, and was represented by St. John in his Apocalypse as the Woman clothed with the Sun, crowned with twelve stars, and with the moon beneath her feet. It is the mystical quintessence of the triad, spirituality, immortality, the queen of heaven.

ד *The porte or government of the easterns, initiation, power, the tetragram, the quaternary, the cubic stone, or its base.*

Hieroglyph, THE EMPEROR, a sovereign whose body represents a right-angled triangle, and his legs a cross, the image of the Athanor of the philosophers.

ה   *Indication, demonstration, instruction, law, symbolism, philosophy, religion.*

Hieroglyph, THE POPE, or grand hierophant. In more modern Tarots this sign is replaced by the image of Jupiter. The grand hierophant, seated between the two pillars of Hermes and of Solomon, makes the sign of esotericism, and leans upon a Cross with three crossbars of triangular form. Two inferior ministers kneel before him. Having above him the capitals of the two pillars and below him the two heads of the assistants, he is thus the center of the quinary, and represents the divine pentagram, giving its complete meaning. As a fact, the pillars are necessity or law, the heads liberty or action. A line may be drawn from each pillar to each head, and two lines from each pillar to each of the two heads. Thus, a square, divided by a cross into four triangles, is obtained, and in the middle of this cross is the grand hierophant, we might almost say like the garden spider in the center of his web, were such a comparison becoming to the things of truth, glory and light.

ו   *Sequence, interlacement, lingam, entanglement, union, embrace, strife, antagonism, combination, equilibrium.*

Hieroglyph, man between Vice and Virtue. Above him shines the sun of truth, and in this sun is Love, bending his bow and threatening Vice with his shaft. In the order of the ten Sephiroth, this symbol corresponds to TIPHERETH – that is, to idealism and beauty. The number six represents the antagonism of the two triads, that is, absolute negation and absolute affirmation. It is therefore the number of toil and liberty, and for this reason it connects also with moral beauty and glory.

THE CHARIOT OF HERMES.
Seventh Key of the Tarot.

ז *Weapon, sword, cherubic sword of fire, the sacred septenary, triumph, royalty, priesthood.*

Hieroglyph, a cubic chariot with four pillars and an azure and starry drapery. In the chariot, between the four pillars, a victor crowned with a circle adorned with three radiant golden pentagrams. Upon his breast are three superposed squares, on his shoulders the urim and thummim of the sovereign sacrificer, represented by the two crescents of the moon in Gedulah and Geburah; in his hand is a scepter surmounted by a globe, square, and triangle; his attitude is proud and tranquil. A double sphinx or two sphinxes joined at the lower parts are harnessed to the chariot; they are pulling in opposite directions, but one is turning his head so that they are looking in the same direction. The sphinx with head turned is black, the other is white. On the square which forms the fore part of the chariot is the Indian lingam surmounted by the flying sphere of the Egyptians. This hieroglyph, which we reproduce exactly, is perhaps the most beautiful and complete of all those that are comprised in the Clavicle of the Tarot.

ח *Balance, attraction and repulsion, life, terror, promise and threat.*

Hieroglyph, JUSTICE with sword and balance.

ט Good, horror of evil, morality, wisdom.

Hieroglyph, a sage leaning on his staff, holding a lamp in front of him and completely enveloped in his cloak. The inscription is THE HERMIT or CAPUCHIN, on account of the hood of his oriental cloak; his true name, however, is PRUDENCE, and he thus completes the four cardinal virtues which seemed imperfect to Court de Gebelin and Etteilla.

פ *Principle, manifestation, praise, manly honor, phallus, virile fecundity, paternal scepter.*

Hieroglyph, THE WHEEL OF FORTUNE, that is to say, the cosmogonical wheel of Ezekiel, with a Hermanubis ascending on the right, a Typhon descending on the left, and a sphinx in equilibrium above, holding a sword between his lion's claws – an admirable symbol, disfigured by Etteilla, who replaced Typhon by a wolf, Hermanubis by a mouse, and the sphinx by an ape, an allegory characteristic of Etteilla's Kabbalah.

ב *The hand in the act of grasping and holding.*

Hieroglyph, STRENGTH, a woman crowned with the vital ∞ closes, quietly and without effort, the jaws of a raging lion.

ל *Example, instruction, public teaching.*

Symbol, a man hanging by one foot, with his hands bound behind his back, so that his body makes a triangle with apex downwards, and his legs cross above the triangle. The gallows is in the form of a Hebrew Tau, and the two uprights are trees, from each of which six branches have been lopped. We have previously explained this symbol of sacrifice and the finished work.

מ *The heaven of Jupiter and Mars, domination and force, new birth, creation and destruction.*

Hieroglyph, DEATH, reaping crowned heads in a meadow where men are growing.

נ *The heaven of the Sun, climates, seasons, motion, changes of life, which is ever new yet ever the same.*

Hieroglyph, TEMPERANCE, an angel with the sign of the sun upon her forehead, and on the breast the square and triangle of the septenary, pours from one chalice into another the two essences which compose the elixir of life.

ס *The heaven of Mercury, occult science, magic, commerce, eloquence, mystery, moral force.*

Hieroglyph, THE DEVIL, the goat of Mendes, or the Baphomet of the Temple, with all his pantheistic attributes. This is the only hieroglyph which was properly understood and correctly interpreted by Etteilla.

ע *The heaven of the Moon, alterations, subversions, changes, failings.*

Hieroglyph, a TOWER struck by lightning, probably that of Babel. Two persons, doubtless Nimrod and his false Prophet or minister, are precipitated from the summit of the ruins. One of the personages in his fall reproduces perfectly the letter *gnaïn*[27].

פ Heaven of the soul, outpourings of thought, moral influence of idea on form, immortality.

Hieroglyph, the BURNING STAR and eternal youth. We have already described this symbol.

צ *The elements, the visible world, reflected light, material forms, symbolism.*

Hieroglyph, the MOON, dew, a crab rising in the water towards land, a dog and wolf barking at the moon and chained to the base of two towers, a path lost in the horizon and sprinkled with blood.

ק *Composites, the head, apex, prince of heaven.*

Hieroglyph, a radiant SUN and two naked children taking hands in a fortified enclosure. Other Tarots substitute a spinner unwinding destinies, and others, again, a naked child mounted on a white horse and displaying a scarlet standard.

---

[27] Ayin.

ר *Vegetative principle, generative virtue of the earth, eternal life.*

Hieroglyph, THE JUDGEMENT. A genius sounds the trumpet and the dead rise from their tombs. These persons, who are living and were dead, are a man, woman and child – the triad of human life.

ש *The sensitive principle, the flesh, eternal life.*

Hieroglyph, the FOOL. A man in the garb of a fool, wandering without aim, burdened with a wallet, full, no doubt, of his follies and vices; his disordered clothes discover his shame; he is being bitten by a tiger and does not know how to escape or defend himself.

ת *The microcosm, the sum of all in all.*

Hieroglyph, KETHER, or the kabbalistic crown, between four mysterious animals. In the middle of the crown is Truth holding a rod in each hand.

Such are the twenty-two keys of the Tarot, which explain all its numbers. Thus, the juggler, or key of the unities, explains the four aces with their quadruple progressive signification in the three worlds and in the first principle. So also, the ace of deniers or of the circle is the soul of the world; the ace of swords is militant intelligence; the ace of cups is loving intelligence; the ace of clubs is creative intelligence; they are also the principles of motion, progress, fecundity and power. Each number, multiplied by a key, gives another number, which, explained in turn by the keys, completes the philosophical and religious revelation contained in each sign. Now, each of the fifty-six cards can be multiplied in turn by the twenty-two keys; a series of combinations thus results, giving all the most astonishing conclusions of revelation and of light. It is a truly philosophical machine, which keeps the mind from going astray while leaving its initiative and liberty; it is mathematics applied to the absolute, the alliance of the positive and the ideal, a lottery of thoughts as exact as numbers, perhaps the simplest and grandest conception of human genius.

The mode of reading the hieroglyphs of the Tarot is to arrange them in a square or triangle, placing equal numbers in antagonism, and conciliating them by the unequal. Four signs invariably express the absolute in a given order, and are explained by a fifth. Hence the solution of all magical questions is the pentagram, and all antinomies are explained by harmonious unity. So arranged, the Tarot is a veritable oracle, and replies to all possible questions with more precision and infallibility than the Android of Albertus Magnus. An imprisoned person with no other book than the Tarot, if he knew how to use it, could in a few years acquire universal knowledge, and would be able to speak on all subjects with unequalled learning and inexhaustible eloquence. In fact, this wheel is the true key to the Oratorical Art and the Grand Art of Raymund Lully; it is the true secret of the transmutation of shadows into light; it is the first and most important of all the arcana of the great work. By means of this universal key of symbolism, all allegories of India, Egypt and Judea are illuminated; the Apocalypse of St. John is a kabbalistic book the sense of which is rigorously indicated by the numbers of the Urim, Thummim, Theraphim and Ephod, which are all resumed and completed by the Tarot; the old sanctuaries contain no longer mysteries, and the significance of the objects of the Hebrew cultus is for the first time comprehensible. Who does not perceive in the golden table, crowned and supported by cherubim, which covered the ark of the covenant, the same symbols as those of the twenty-first Tarot key? The ark was a hieroglyphical synthesis of the whole kabbalistic dogma; it included the *jod* or blossoming staff of Aaron, the *he* or cup, the gomor containing the manna, the two tables of the law – an analogous symbol to that of the sword of justice – and the manna kept in the gomor, four objects which interpret wonderfully the letters of the divine tetragram. Gaffarel has learnedly proved that the cherubim, or cherubs of the ark, were in the likeness of bulls, but what he did not know was that, instead of two, there were four – two at each end, as the text expressly says – though it has been misconstrued for the most part by commentators. The eighteenth and nineteenth verses of the twenty-fifth chapter of Exodus should read thus: "And thou shalt make two bulls or sphinxes of beaten gold on each side of the oracle. And thou shalt make the one looking this way and the second that way.[28]" The cherubim or sphinxes were, in fact, coupled by twos on each side of the ark, and their heads were turned to the four corners of the mercy-seat, which they covered with their wings rounded archwise, thus overshadowing the crown of the golden table, which they sustained upon their shoulders, facing one another at the openings and looking at the propitiatory, as shown in the figure below. The ark, moreover, had three parts or stages, representing Atziluth, Yetzirah and Briah – the three worlds of the kabbalah: the base of the coffer, to which were fitted the four rings of two levers analogous to

---

[28] "18 And thou shalt make two cherubims of gold, of beaten work shalt thou make them, in the two ends of the mercy seat. 19 And make one cherub on the one end, and the other cherub on the other end: even of the mercy seat shall ye make the cherubims on the two ends thereof." Book of Exodus, King James Version

the pillars of the temple, JAKIN and BOHAS; the body of the coffer, on which the sphinxes appeared in relief; and the cover, overshadowed by the wings. The base represented the kingdom of salt, to use the terminology of the adepts of Hermes; the coffer, the realm of mercury or azoth; and the cover, the realm of sulphur or of fire. The other objects of the cultus were not less allegorical, but would require a special treatise to describe and explain them.

Saint-Martin, in his Natural Table of the Correspondences between God, Man and the Universe, followed, as we have said, the division of the Tarot, giving an extended mystical commentary upon the twenty-two keys, but he carefully refrained from stating whence he derived his plan, and from revealing the hieroglyphics on which he commented. Postel shewed similar discretion, naming the Tarot only in the diagram of the key to his arcana, and referring to it in the rest of his book under the title of the Genesis of Enoch. The personage Enoch, author of the primeval sacred book, is in effect identical with that of Thoth among the Egyptians, Cadmus among the Phœnicians, and

Palamedes among the Greeks. We have obtained in an extraordinary manner a sixteenth-century medal, which is a key of the Tarot. We scarcely know whether to state that this medal, and the place where it was deposited, were shown us in dream by the divine Paracelsus; in any case, the medal is in our possession. On one side it depicts the juggler in a German costume of the sixteenth century, holding his girdle with one hand, and with the other a Pentagram. On a table in front of him, between an open book and a closed purse, are ten deniers or talismans, arranged in two lines of three each and a square of four; the feet of the table form two ה and those of the juggler two inverted ר. The obverse side of the medal contains the letters of the alphabet, arranged in a magical square, as follows:

It will be observed that this alphabet has only twenty-two letters, the V and N being duplicated, and that it is arranged in four quinaries, with a quaternary for base and key. The four final letters are two combinations of the duad and the triad, and, read kabbalistically, they form the word AZOTH, by rendering to the shapes of the letters their value in primitive Hebrew, taking N for א, Z as it is in Latin, V for the Hebrew ו vau, which is pronounced O between two vowels, or letters having the value of vowels, and X for the primitive tau, which had precisely the same figure. The entire Tarot is thus explained in this wonderful medal, which is worthy of Paracelsus, and we hold it at the disposal of the curious. The letters arranged by four times five are summed by the word אZוה, analogous to those of יהוה and INRI, and containing all the mysteries of the Kabbalah.

The Book of the Tarot, being of such high scientific importance, it is desirable that it should not be further altered. We have examined the collection of ancient Tarots preserved in the Imperial Library, and have thus collected all the hieroglyphs, of which we have given a description. An important work still remains to be done – the publication of a really complete and well-executed exemplar. We shall, perhaps, undertake the task.

Vestiges of the Tarot are found among all nations. As we have said, the Italian is, perhaps, the most faithful and best preserved, but it may be further perfected by precious indications derived from the Spanish varieties. The two of cups, for example, in the Naïbi is completely Egyptian, showing two archaic vases with ibis handles, superposed on a cow. A unicorn is represented in the middle of the four of deniers; the three of cups exhibits the figure of Isis issuing from a vase, while two ibises issue from two other vases, one with a crown for the goddess, and one holding a lotus, which he seems to be offering for her acceptance. The four aces bear the image of the hieratic and sacred serpent, while in some specimens the seal of Solomon is placed at the center of the four of deniers, instead of the symbolical unicorn. The German Tarots have suffered great alteration, and scarcely do more than preserve the number of the keys, which are crowded with grotesque or pantagruelian figures. We have a Chinese Tarot before us, and the Imperial Library contains samples of others that are similar. M. Paul Boiteau, in his remarkable work on playing-cards, has given some admirably executed specimens. The Chinese Tarot preserves several primeval emblems; the deniers and swords are plainly distinguishable, but it would be less easy to discover the cups and clubs.

It was at the epoch of the Gnostic and Manichæan heresies that the Tarot must have been lost to the Church, at which time also the meaning of the divine Apocalypse perished. It was understood no longer that the seven seals of this kabbalistic book are seven pantacles, the representation of which we give *(The Seven Seals of St. John)*, and that these pantacles are explained by the analogies of the numbers, characters, and figures of the Tarot. Thus, the universal tradition of the one religion was a moment broken, darkness or doubt spread over the whole earth, and it seemed, in the eyes of ignorance, that true Catholicism, the universal revelation, had briefly disappeared. The explanation of the book of St. John by the characters of the Kabbalah will be an entirely new revelation, though foreseen by several distinguished magi, one among whom, M. Augustin Chaho, thus expresses himself:

"The poem of the Apocalypse presupposes in the young evangelist a complete system and traditions individually developed by himself. It is written in the form of a vision, and binds in a brilliant framework of poetry the whole erudition, the whole thought of African civilization. An inspired bard, the author touches upon a series of ruling events; he draws in bold outlines the history of society from cataclysm to cataclysm, and even further still. The truths which he reveals are prophecies brought from far and wide, of which he is the resounding echo. He is the voice which cries, the voice which chants the harmonies of the desert, and prepares paths for the light. His speech peals forth with mastery and compels faith, for he carries among savage nations the oracles of *Iao*, and unveils Him Who is the First-Born of the Sun for the admiration of civilizations to come. The theory of the four ages is found in the Apocalypse, as it is also in the books of Zoroaster and in the Bible. The gradual reconstruction of primeval federations, of the reign of God among peoples emancipated from the yoke of tyrants and the bonds of error, are clearly foretold for the end of the fourth age, and the renovation of the cataclysm, exhibited at first from afar, even unto the consummation of time. The description of the cataclysm and its duration; the new world emerging from the waves, and spreading in all its beauty under heaven; the great serpent, bound for a time by an angel in the depths of the abyss; finally, the dawn of that age to come, prophesied by the Word, who appeared to the apostle at the beginning of his poem: 'His head and his hairs were white like wool, as white as snow, and his eyes were as a flame of fire; and his feet like unto fine brass, as if they burned in a furnace; and his voice as the sound of many waters. And he had in his right hand seven stars: and out of his mouth went a sharp two-edged sword: and his countenance was as the sun shineth in his strength.' Such is Ormuz, Osiris, Chourien, the Lamb, the Christ, the Ancient of Days, the man of the time and the river celebrated by Daniel. He is the first and the last, who was, who must be, alpha and omega, beginning and end. He holds the key of mysteries in his hands; he opens the great abyss of the central fire, where death sleeps beneath his canopy of darkness, where sleeps the great serpent awaiting the wakening of the ages."

APOCALYPTIC KEY.
The Seven Seals of St John.

The author connects this sublime allegory of St. John with that of Daniel, wherein the four forms of the sphinx are applied to the chief periods of history, where the Man-Sun, the Word-Light, consoles and instructs the seer.

"The prophet Daniel beholds a sea tossed by the four winds of heaven, and beasts differing one from another come out of the depths of the ocean. The empire of all things on earth was given them for a time, two times, and the dividing of time. They are four who so came forth. The first beast, symbol of the solar race of seers, comes from the region of Africa, resembling a lion and having eagle's wings; the heart of a man was given it. The second beast, emblem of the northern conquerors, who reigned by iron during the second age, was like unto a bear; it had three ribs in the mouth of it between the teeth of it, images of the three great conquering families, and they said unto it: Arise, devour much flesh. After the apparition of the fourth beast, there were thrones raised up, and the Ancient of Days, the Christ of seers, the Lamb of the first age, was manifested. His garment was of dazzling whiteness, his head radiant; his throne, whence came forth living flames, was borne upon burning wheels; a flame of swift fire shone in his countenance; legions of angels or stars sparkled round him. The judgement was set, the allegorical books were opened. The new Christ came with the clouds of heaven and came to the Ancient of Days; there were given him power, honor, and a kingdom over all peoples, tribes, and tongues. Then Daniel came near unto one of them that stood by, and asked him the truth of all this. And it was answered him that the four beasts were four powers which should reign successively over the earth." M. Chaho proceeds to explain a variety of images, strikingly analogous, which are found in almost all sacred books. His observations at this point are worthy of remark.

"In every primitive logos, the parallel between physical correspondences and moral relations is established on the same roots. Each word carries its material and sensible definition, and this living language is as perfect and true as it is simple and natural in man the creator. Let the seer express by the same word, slightly modified, the sun, day, light, truth, and applying the same epithet to a white sun and to a lamb, let him say, *Lamb* or *Christ*, instead of *sun*, and *sun* instead of *truth, light, civilization*, and there is no allegory, but there are true correspondences seized and expressed by inspiration. But when the children of night say in their incoherent and barbarous dialect, *sun, day, light, truth, lamb*, the wise correspondence so clearly expressed by the primitive logos becomes effaced and disappears, and, by simple translation, the lamb and the sun become allegorical beings, symbols. Remark, in effect, that the word *allegory* itself signifies in Celtic definition, *change of discourse, translation*. The observation we have made applies exactly to all barbarous cosmogonical language. Seers made use of the same inspired radical to express *nourishment* and *instruction*. Is not the science of truth the nourishment of the soul? Thus, the scroll of papyrus, or the book, eaten by the prophet Ezekiel; the little volume which the angel gave as food to the author of the Apocalypse; the festivities of the magical palace of Asgard, to which Gangler was invited by Har the Sublime; the miraculous multiplication of seven small loaves narrated by the Evangelists of the Nazarene; the living bread which Jesus-Sun gave his disciples to eat, saying, 'This is my body'; and a host of similar occurrences, are a repetition of the same allegory: the life of souls who are nourished by truth – truth, which multiplies without ever diminishing, but, on the contrary, increases in the measure that it nourishes.

"Exalted by a noble sentiment of patriotism, dazzled by the idea of an immense revolution, let a revealer of hidden things come forward and seek to popularize the discoveries of science among gross and ignorant men, destitute of the most simple elementary notions; let him say, for example, that the earth revolves, and that it is shaped like an egg; what resource has the barbarian who hears him except *to believe*? Is it not plain that every proposition of this nature becomes for him a dogma from on high, an article of *faith*? And is not the veil of a wise allegory sufficient to make it a mythos? In the schools of seers, the terrestrial globe was represented by an egg of pasteboard or painted wood, and when young children were asked, 'What is this egg?' they answered, 'It is the earth.' Those older children, the barbarians, hearing this, repeated after the little children of the seers: 'The world is an egg.' But they understood thereby the physical, material world, while the seers meant the geographical, ideal, image-world, created by mind and the logos. As a fact, the priests of Egypt represented mind, intelligence, Kneph, with an egg placed upon his lips, to express clearly that the egg was here as only a comparison, an image, a mode of speech. Choumountou, the philosopher of the Ezour-Vedam, explains after the same manner to the fanatic Biache what must be understood by the golden egg of Brahma."

We must not wholly despair of a period which still concerns itself with these serious and reasonable researches; we have cited these pages of M. Chaho with great mental satisfaction and profound sympathy. Here is no longer the negative and desolating criticism of Dupuis and Volney, but tendency towards one faith and one worship connecting all the future with all the past; it is the exoneration of all great men accused falsely of superstition and idolatry; it, is, finally, the justification of God Himself, that sun of intelligences who is never veiled for just souls and pure hearts.

"Great and pre-eminent is the seer, the initiate, the elect of nature and of supreme reason," cries the author once more, in concluding what we have just cited. "His alone is that faculty of imitation which is the principle of his perfection, while its inspirations, swift as a lightning flash, direct creations and discoveries. His alone is a perfect Word of conformity, propriety, flexibility, wealth, creating harmony of thought by physical reaction – of thought, whereof the perceptions, still independent of language, ever reflect nature exactly reproduced in his impressions, well-judged and well expressed in its correspondences. His alone is light, science, truth, because imagination, confined to its passive secondary part, never governs reason, the natural logic which results from the comparison of ideas; which come into being, extend in the same proportion as his needs, and because the circle of his knowledge enlarges thus by degrees without intermixture of false judgements and errors. His alone is a light infinitely progressive, because the rapid multiplication of population, after terrestrial renovations, composes in a few centuries a new society in all the imaginable moral and political correspondences of its destiny; and we might add, his alone is absolute light. The man of our time is immutable in himself; he changes no more than nature, in which he is rooted. The social conditions which surround him alone determine the degree of his perfection, of which the bounds are virtue, holiness of man, and his happiness in the law. "

After such elucidations, will anyone ask the utility of the occult sciences? Will they treat with the disdain of mysticism and illuminism these living mathematics, these proportions of ideas and forms, this revelation permanent in the universal reason, this emancipation of mind, this immutable basis provided for faith, this omnipotence revealed to will? Children in search of illusions, are you disappointed because we offer you marvels? Once a man said to us, "Raise up the devil, and I will believe in you." We answered, "You ask too little; we will not make the devil appear but vanish from the whole world; we will chase him from your dreams!" The devil is ignorance, darkness, chaotic thought, deformity. Awake, sleeper of the middle ages! See you not that it is day? See you not the light of God filling all nature? Where now will the destroyed prince of perdition dare to show himself?

It remains for us to state our conclusions and to define the end and application of this work in the religious and philosophical order, and in the order of positive and material realizations. As regards the religious order, we have demonstrated that the practices of religious worships cannot be indifferent, that the magic of religions is in their rites, that their moral force is in the triadic hierarchy, and that the base, principle and synthesis of the hierarchy is unity. We have demonstrated the universal unity and orthodoxy of dogma, clothed successively with various allegorical veils, and we have followed the truth saved by Moses from profanation in Egypt, preserved in the kabbalah of the prophets, emancipated by the Christian school from the slavery of the pharisees, attracting all the poetic and generous aspirations of Greek and Roman civilization, protesting against a new pharisaism more corrupt than the first, with the great saints of the middle ages and the bold thinkers of the Renaissance. We have exhibited, I say, that truth always universal, always living, alone conciliating reason and faith, science and submission; the truth of being demonstrated by being, of harmony demonstrated by harmony, of reason manifested by reason. By revealing for the first time to the world the mysteries of magic we have not sought to revive practices entombed beneath the ruins of ancient civilizations, but would say to humanity in our day that it is also called to create itself immortal and omnipotent by its works. Liberty does not offer itself, it must be seized, says a modern writer: it is the same with science, for which reason to divulge absolute truth is never useful to the vulgar. But at an epoch when the sanctuary has been devastated and has fallen into ruins, because its key has been thrown over the hedges to the profit of no one, I have deemed it my duty to pick up that key, and I offer it to him who can take it; in his turn he will be doctor of the nations and liberator of the world. Fables and leading-strings are needed, and will always be needed by children, but it is not necessary that those who hold the leading-strings should also be children, lending a ready ear to fables. Let the most absolute science, let the highest reason, become the possession of the chiefs of the people; let the priestly art and the royal art take up once more the double scepter of antique initiations, and the world will re-issue from chaos. Burn no more holy images, destroy no more temples; temples and images are necessary for man; but drive out the merchants from the house of prayer; let the blind no longer be leaders of the blind; reconstruct the hierarchy of intelligence and holiness, and recognize only those who know as the teachers of those who believe. Our book is catholic, and if the revelations it contains are likely to alarm the conscience of the simple, we are consoled by the thought that they will not read them. We write for unprejudiced men, and have no wish to flatter irreligion any more than fanaticism. If there be anything essentially free and inviolable in the world, it is belief. By science and persuasion, we must endeavor to lead bewrayed imaginations from the absurd, but it would be investing their errors with all the dignity and truth of the martyr to either threaten or constrain them.

Faith is nothing but superstition and folly if it have not reason for its basis, and we cannot suppose that which we do not know except by analogy with what we know. To define what we are unacquainted with is presumptuous ignorance; to affirm positively what one does not know is to lie. So is faith an aspiration and a desire. So be it; I desire it to be so; such is the last word of all professions of faith. Faith, hope, and charity are three such inseparable sisters that they may be taken one for another. Thus, in religion, universal and hierarchic orthodoxy, restoration of temples in all their splendor, re-establishment of all ceremonies in their primitive pomp, hierarchic instruction of symbols, mysteries, miracles, legends for children, light for grown men who will beware of scandalizing little ones in the simplicity of their faith; this in religion is our whole utopia, and it is also the desire and need of humanity.

Coming now to philosophy, our own is that of realism and positivism. Being is by reason of the being of which no one doubts. All exists for us by science. To know is to be. Science and its object become identified in the intellectual life of him who knows. To doubt is to be ignorant. Now, a thing of which we are ignorant does not as yet exist for us. To live intellectually is to learn. Being develops and amplifies by science. The first conquest of science, and the first result of the exact sciences, is the sentiment of reason. The laws of nature are algebraic. Thus, the sole reasonable faith is the adhesion of the student to theorems, the entire essential justice of which lies outside his knowledge, though its applications and results are sufficiently demonstrated to his mind. Thus, the true philosopher believes in what is, and does not admit *à posteriori* that all is reasonable. But no more charlatanism in philosophy, no more empiricism, no more system! The study of being and its compared realities! A metaphysic of nature! Then away with mysticism! No more dreams in philosophy; philosophy is not a poesy, but the pure mathematics of realities, physical and moral. Leave unto religion the freedom of its infinite aspirations, and let it leave in turn to science the exact conclusions of absolute experimentalism.

Man is the son of his works; he is what he wills to be; he is the image of the God he makes; he is the realization of his ideal. Should his ideal want basis, the whole edifice of his immortality collapses. Philosophy is not the ideal, but it serves as a foundation for the ideal. The known is for us the measure of the unknown; by the visible we appreciate the invisible; sensations are to thoughts even as thoughts to aspirations. Science is a celestial trigonometry: one of the sides of the absolute triangle is the nature which is submitted to our investigations; the second is our soul, which embraces and reflects nature; the third is the absolute, in which our soul enlarges. No more atheism possible henceforward, for we no longer pretend to define God. God is for us the most perfect and best of intelligent beings, and the ascending hierarchy of beings sufficiently demonstrates his existence. Do not let us ask for more, but, to be ever understanding him better, let us grow perfect by ascending towards him. No more ideology; being is being, and cannot perfectionize save according to real laws of being. Observe, and do not prejudge; exercise our faculties, do not falsify them; enlarge the domain of life in life; behold truth in truth! Everything is possible to him who wills only what is true! Rest in nature, study, know, then dare; dare to will, dare to act, and be silent! No more hatred of anyone.

Everyone reaps what he sows. The consequence of works is fatal, and to judge and chastise the wicked is for the supreme reason. He who enters into a blind alley must retrace his steps or be broken. Warn him gently, if he can still hear you, but human liberty must take its course. We are not the judges of one another. Life is a battlefield. Do not pause in the fighting on account of those who fall, but avoid trampling them. Then comes the victory, and the wounded on both sides, become brothers by suffering and before humanity, will meet in the ambulances of the conquerors.

Such are the consequences of the philosophical dogma of Hermes; such has been from all time the ethic of true adepts; such is the philosophy of the Rosicrucian inheritors of all the ancient wisdoms; such is the secret doctrine of those associations that are treated as subversive of the public order, and have ever been accused of conspiring against thrones and altars. The true adept, far from disturbing the public order, is its firmest supporter. He has too great a respect for liberty to desire anarchy; child of the light, he loves harmony, and knows that darkness begets confusion. He accepts everything that is and denies only what is not. He wills true religion, practical, universal, full of faith, palpable, realized in all life; he wills it to have a wise and powerful priesthood, surrounded by all the virtues and all the prestige of faith. He wills the universal orthodoxy, the absolute, hierarchic, apostolic, sacramental, incontestable, and uncontested catholicity. He wills an experimental philosophy, real, mathematical, modest in its conclusions, untiring in its researches, scientific in its progress. Who, therefore, can be against us if God and reason are with us? Does it matter if man prejudge and slander us? Our entire justification is in our thoughts and our works. We come not, like Œdipus, to destroy the sphinx of symbolism; we seek, on the contrary, to resuscitate it. The sphinx devours only blind interpreters; and he who slays it has not known how to divine it properly; it must be subdued, enchained, and compelled to follow us. The sphinx is the living palladium of humanity, it is the conquest of the King of Thebes; it would have been the salvation of Œdipus, had Œdipus completely divined its enigma!

In the positive and material order, what must be concluded from this work? Is magic a force which science may abandon to the boldest and wickedest? Is it a cheat and falsehood of those who are skilled in fascinating the ignorant and feeble? Is the philosophical mercury the exploitation of credulity by address? Those who have understand us know already how to answer these questions. In these days, magic can be no longer the art of fascinations and illusions; those only who wish to be deceived can be deceived now. But the narrow and rash incredulity of the last century is denied in totality by nature herself. We are environed with prophecies and miracles; doubt once unwisely denied them; now, science explains them. No, Monsieur le Comte de Mirville, a destroyed spirit is not allowed to disturb the empire of God! No, things unknown cannot be explained by things impossible! No, invisible beings are not permitted to deceive, torment, seduce and even kill the living creatures of God, men, already so ignorant, and scarce able to combat their own delusions! Those who told you all this in your childhood, Monsieur le Comte, have deceived you, and if you were child enough once to listen to them, be man enough now to disbelieve them. Man is himself the creator of his heaven and hell, and there are no demons except our own follies. Minds chastised by truth are corrected by that chastisement, and dream no more of disturbing the world. If Satan exists, he can be only the most unfortunate, most ignorant, most humiliated, and most impotent of beings. The existence of a universal agent of life, of a living fire, of an astral light, is demonstrated by facts. Magnetism enables us to understand today the miracles of old magic; the facts of second sight, aspirations, sudden cures, thought-reading, are now admitted and familiar things, even among our children. But the tradition of the ancients has been lost, discoveries have been regarded as new, the last word is sought about observed phenomena, minds grow excited over meaningless manifestations, fascinations are experienced without being understood. We say, therefore, to table-turners: These prodigies are not novel; you can perform even greater wonders if you study the laws of nature. And what will follow a new acquaintance with these powers? A new career opened to the activity and intelligence of man, the battle of life reorganized with arms more perfect, and the possibility restored to the flower of intelligence of once more becoming the masters of all destinies, by providing true priests and great kings for the world to come!

HERE ENDS THE RITUAL OF TRANSCENDENT MAGIC.

Éliphas Lévi

SUPPLEMENT TO THE RITUAL

THE NUCTEMERON OF APOLLONIUS OF TYANA

The Greek text was first published after an ancient manuscript, by Gilbert Gautrinus, in De Vita et Morte Moysis, Lib. III, p. 206; and subsequently reproduced by Laurent Moshemius in his Sacred and Historico-Critical Observations. Amsterdam, 1721. Translated and interpreted for the first time by Éliphas Lévi.

NUCTEMERON signifies the day of the night or the night illuminated by day. It is analogous to the "Light Issuing from Darkness," which is the title of a well-known Hermetic work. It may also be translated THE LIGHT OF OCCULTISM. This monument of transcendent Assyrian magic is sufficiently curious to make it superfluous to enlarge on its importance. We have not merely evoked Apollonius, we have possibly resuscitated him.

THE NUCTEMERON

The First Hour.

In unity, the demons chant the praises of God; they lose their malice and fury.

The Second Hour.

By the duad, the Zodiacal fish chant the praises of God; the fiery serpents interlace about the caduceus, and the lightning becomes harmonious.

The Third Hour.

The serpents of the Hermetic caduceus entwine three times; Cerberus opens his triple jaw, and fire chants the praises of God with the three tongues of the lightning.

The Fourth Hour.

At the fourth hour the soul revisits the tombs; the magical lamps are lighted at the four corners of the circle; it is the time of enchantments and illusions.

The Fifth Hour.

The voice of the great waters celebrates the God of the heavenly spheres.

The Sixth Hour.

The spirit believes itself immovable; it beholds the infernal monsters swarm down upon it, and does not fear.

The Seventh Hour.

A fire, which imparts life to all animated beings, is directed by the will of pure men. The initiate stretches forth his hand, and pains are assuaged.

The Eighth Hour.

The stars utter speech to one another; the soul of the suns corresponds with the exhalation of the flowers; chains of harmony create correspondence between all-natural things.

The Ninth Hour.

The number which must not be divulged.

The Tenth Hour.

The key of the astronomical cycle and of the circular movement of human life.

The Eleventh Hour.

The wings of the genii move with a mysterious and deep murmur; they fly from sphere to sphere, and bear the messages of God from world to world.

The Twelfth Hour.

The works of the light eternal are fulfilled by fire.

EXPLANATION

THESE twelve symbolical hours, analogous to the signs of the magical Zodiac and to the allegorical labors of Hercules, represent the schedule of the works of initiation. It is necessary therefore (1) To overcome evil passions, and, according the expression of the wise Hierophant, compel the demons themselves to praise God. (2) To study the balanced forces of nature, and know how harmony results from the analogy of contraries; to know also the great magical agent and the twofold polarization of the universal light. (3) To gain initiation into the triadic principle of all theogonies and all religious symbols. (4) To know how to overcome all phantoms of imagination, and triumph over all illusions. (5) To understand after what manner universal harmony is produced in the center of the four elementary forces. (6) To become inaccessible to fear. (7) To practice the direction of the magnetic light. (8) To learn prevision of efforts by the calculus of the balance of causes. (9) To understand the hierarchy of instruction, to respect the mysteries of dogma, and to keep silence in presence of the profane. (10) To study astronomy exhaustively. (11) To become initiated by analogy into the laws of universal life and intelligence. (12) To fulfil the great works of nature by direction of the light.

Here follow the names and attributions of the genii who preside over the twelve hours of the Nuctemeron. By these genii the ancient hierophants understood neither angels nor demons, but moral forces or personified virtues.

Genii of the First Hour.

PAPUS, physician. SINBUCK, judge. RASPHUIA, necromancer. ZAHUN, genius of scandal. HEIGLOT, genius of snowstorms. MIZKUN, genius of amulets. HAVEN, genius of dignity.

Explanation.

We must become the *physician* and *judge* of ourselves in order to vanquish the witchcrafts of the *necromancer*, conjure and contemn the genius of *scandal*, triumph over the opinion which freezes all enthusiasms, and confounds everything in the same cold pallor, like the *genius of the snowstorms*; know, finally, the virtue of signs so as to enchain the *genius of amulets* that we may reach the *dignity* of the magus.

Genii of the Second Hour.

SISERA, genius of desire. TORVATUS, genius of discord. NITIBUS, genius of the stars. HIZARBIN, genius of the seas. SACHLUPH, genius of plants. BAGLIS, genius of measure and balance. LABEZERIN, genius of success.

Explanation.

We must learn how to will and thus transform the *genius of desire* into power; the hindrance to will is the *genius of discord*, who is bound by the science of harmony. Harmony is the *genius of the stars and of the seas*; we must study the virtue of *plants,* and understand the laws of the *balance of measure* in order to attain *success*.

Genii of the Third Hour.

HAHABI, genius of fear. PHLOGABITUS, genius of adornments. EIRNEUS, destroying genius of idols. MASCARUN, genius of death. ZAROBI, genius of precipices. BUTATAR, genius of calculations. CAHOR, genius of deception.

Explanation.

When you have conquered the *genius of fear* by the growing force of your will, you will know that dogmas are the sacred *adornments* of truth unknown to the vulgar; but you will cast down all *idols* in your intelligence; you will bind the *genius of death*; you will fathom all *precipices* and subject the infinite itself to the ratio of your calculations; and thus you will ever escape the ambushes of the *genius of deception*.

Genii of the Fourth Hour.

PHALGUS, genius of judgment. THAGRINUS, genius of confusion. EISTIBUS, genius of divination. PHARZUPH, genius of fornication. SISLAU, genius of poisons. SCHIEKRON, genius of bestial love. ACLAHAYR, genius of sport.

Explanation.

The power of the magus is in his *judgment*, which enables him to avoid the *confusion* consequent on antinomy and the antagonism of principles; he practices the *divination* of the sages, but he despises the illusions of enchanters who are the slaves of *fornication*, artists in poisons, ministers of *bestial love*; in this way he is victorious over fatality, which is the *genius of sport*.

Genii of the Fifth Hour.

ZEIRNA, genius of infirmities. TABLIBIK, genius of fascination. TACRITAU, genius of goëtic magic. SUPHLATUS, genius of the dust. SAIR, genius of the stibium[29] of the sages. BARCUS, genius of the quintessence. CAMAYSAR, genius of the marriage of contraries.

Explanation.

Triumphing over human *infirmities*, the magus is no longer the sport of *fascination*; he tramples on the vain and dangerous practices of *goëtic magic*, the whole power of which is but *dust* driven before the wind; but he possesses the *stibium* of the sages; he is armed with all the creative powers of the *quintessence*; and he produces at will the harmony which results from the analogy and the *marriage of contraries*.

---

[29] Antimony.

Genii of the Sixth Hour.

TABRIS, genius of free will. SUSABO, genius of voyages. EIRNILUS, genius of fruits. NITIKA, genius of precious stones. HAATAN, genius who conceals treasures. HATIPHAS, genius of attire. ZAREN, avenging genius.

Explanation.

The magus is free; he is the occult king of the earth, and he traverses it as one passing through his own domain. In his voyages he becomes acquainted with the juices of plants and *fruits*, and with the virtues of *precious stones*; he compels the *genius who conceals the treasures* of nature to deliver him all his secrets; he thus penetrates the mysteries of form; understands the *vestures* of earth and speech; and if he be misconstrued, if the nations are inhospitable towards him, if he pass doing good but receiving outrages, then is he ever followed by the *avenging genius*.

Genii of the Seventh Hour.

SIALUL, genius of prosperity. SABRUS, sustaining genius. LIBRABIS, genius of hidden gold. MIZGITARI, genius of eagles. CAUSUB, serpent-charming genius. SALILUS, genius who sets doors open. JAZER, genius who compels love.

Explanation.

The septenary expresses the victory of the magus; who gives *prosperity* to men and nations; who *sustains* them by his sublime instructions; he broods like the *eagle*; who directs the currents of the astral fire, represented by *serpents*; the gates of all sanctuaries are open to him, and in him all souls who yearn for truth repose their trust; he is beautiful in his moral grandeur; and everywhere does he take with him that genius by the power of which we obtain *love*.

Genii of the Eighth Hour.

NANTUR, genius of writing. TOGLAS, genius of treasures. ZALBURIS, genius of therapeutics. ALPHUN, genius of doves. TUKIPHAT, genius of the schamir. ZIZUPH, genius of mysteries. CUNIALI, genius of association.

Explanation.

These are the genii who obey the true magus; the *doves* represent religious ideas; the *schamir* is an allegorical diamond, which in magical traditions represents the stone of the sages, or that force which nothing can resist, because it is based on truth. The Arabs still say that the schamir, originally given to Adam and lost by him after his fall, was recovered by Enoch and possessed by Zoroaster; and that Solomon subsequently received it from an angel when he entreated wisdom from God. By means of this magical diamond, Solomon himself dressed the stones of the temple merely by touching them with the schamir.

Genii of the Ninth Hour.

RISNUCH, genius of agriculture. SUCLAGUS, genius of fire. KIRTABUS, genius of languages. SABLIL, genius who discovers thieves. SCHACHLIL, genius of the sun's rays. COLOPATIRON, genius who sets prisons open. ZEFFAR, genius of irrevocable choice.

Explanation.

This number, says Apollonius, must be passed over in silence, because it contains the great secrets of the initiate, the power which *fructifies the earth*, the mysteries of *secret fire*, the universal key of languages, the second sight from which *evil-doers* cannot remain concealed. The great laws of equilibrium and of luminous motion represented by the four animals of the Kabbalah, and in Greek mythology by the four horses of the sun, the key of the emancipation of bodies and souls, opening all *prisons*, and that power of eternal *choice* which completes the creation of man and establishes him in immortality.

Genii of the Tenth Hour.

SEZARBIL, devil or hostile genius. AZEUPH, destroyer of children. ARMILUS, genius of cupidity. KATARIS, genius of dogs or of the profane. RAZANIL, genius of the onyx. BUCAPHI, genius of stryges. MASTHO, genius of delusive appearances.

Explanation.

Numbers end with nine, and the distinctive sign of the denary is zero, itself without value, added to unity. The genii of the tenth hour represent all which, being nothing in itself, receives great power from opinion, and can suffer consequently the omnipotence of the sage. We tread here on hot earth, and we must be permitted to omit elucidations to the profane as to *the devil*, who is their master, or the *destroyer of children*, who is their love, or the *cupidity*, which is their god, or the *dogs*, to which we do not compare them, or to the *onyx*, which they possess not, or to the *stryges*, who are their courtesans, or to the *false appearances* which they take for truth.

Genii of the Eleventh Hour.

ÆGLUN, genius of the lightning. ZUPHLAS, genius of forests. PHALDOR, genius of oracles. ROSABIS, genius of metals. ADJUCHAS, genius of rocks. ZOPHAS, genius of pantacles. HALACHO, genius of sympathies.

Explanation.

The *lightning* obeys man; it becomes the vehicle of his will, the instrument of his power, the light of his torches; the oaks of the sacred *forests* utter *oracles*: *metals* change and transmute into gold, or become talismans; *rocks* move from their foundation, and drawn by the lyre of the grand hierophant, touched by the mysterious schamir, transform into temples and palaces; dogmas evolve; symbols represented by *pantacles* become efficacious; minds are enchained by powerful *sympathies*, and obey the laws of family and friendship.

Genii of the Twelfth Hour.

TARAB, genius of extortion. MISRAN, genius of persecution. LABUS, genius of inquisition. KALAB, genius of sacred vessels. HAHAB, genius of royal tables. MARNES, genius of the discernment of spirits. SELLEN, genius of the favor of the great.

Explanation.

Here now is the fate which the magi must expect, and how their sacrifice must be consummated; for after the conquest of life, they must know how to immolate themselves in order to be reborn immortal. They will suffer extortion; gold, pleasure, vengeance will be required of them, and if they fail to satisfy vulgar cupidities they will be the objects of persecution and inquisition; yet the sacred vessels are not profaned; they are made for royal tables, that is, for the feasts of the understanding. By the discernment of spirits, they will know how to protect themselves from the favor of the great, and they will remain invincible in their strength and in their liberty.

THE NUCTEMERON ACCORDING TO THE HEBREWS

Extracted from the ancient Talmud termed Mischna
By the Jews

The Nuctemeron of Apollonius, borrowed from Greek theurgy, completed and explained by the Assyrian hierarchy of genii, perfectly corresponds to the philosophy of numbers as we find it expounded in the most curious pages of the Talmud. Thus, the Pythagoreans go back further than Pythagoras; thus, Genesis is a magnificent allegory, which, under the form of a narrative, conceals the secrets not only of the creation achieved of old, but of permanent and universal creation, the eternal generation of beings. We read as follows in the Talmud: "God hath stretched out the heaven like a tabernacle; He hath spread the world like a table richly dight; and He hath created man as if He invited a guest." Listen now to the words of the King Schlomoh: "The divine Chocmah, Wisdom, the Bride of God, hath built a house unto herself, and hath dressed two pillars. She hath immolated her victims, she hath mingled her wine, she hath spread the table, and she hath dispatched her servitors." This Wisdom, who builds her house according to a regular and numerical architecture, is that exact science which rules the works of God. It is His compass and His square. The seven pillars are the seven typical and primordial days. The victims are natural forces which are propagated by undergoing a species of death.

Mingled wine is the universal fluid, the table is the world with the waters full of fishes. The servants of Chocmah are the souls of Adam and of Chavah (Eve). The earth of which Adam was formed was taken from the entire mass of the world. His head is Israël, his body the empire of Babylon, and his limbs are the other nations of the earth. (Here manifest the aspirations of the initiates of Moses towards a universal oriental kingdom.) Now, there are twelve hours in the day of man's creation.

First Hour.

God combines the scattered fragments of earth; he kneads them together, and forms one mass, which it is his will to animate.

Explanation.

Man is the synthesis of the created world; in him recurs the creative unity; he is made in the image and likeness of God.

Second Hour.

God designs the form of the body; he separates it into two sections, so that the organs may be double, for all force and all life result from two, and it is thus the Elohim made all things.

Explanation.

Everything lives by movement, everything is maintained by equilibrium, and harmony results from the analogy of contraries; this law is the form of forms, the first manifestation of the activity and fecundity of God.

Third Hour.

The limbs of man, obeying the law of life, manifest of themselves and are completed by the generative organ, which is composed of one and two, figure of the triadic number.

Explanation.

The triad issues spontaneously from the duad; the movement which produces two also produces three; three is the key of numbers, for it is the first numerical synthesis; in geometry it is the triangle, the first complete and enclosed figure, generatrix of an infinity of triangles, whether like or unlike.

Fourth Hour.

God breathes upon the face of man and imparts to him a soul.

Explanation.

The tetrad, which geometrically gives the cross and the square, is the perfect number; now, it is in perfection of form that the intelligent soul manifests; according to this revelation of the Mischna, the child would not become animated in the mother's womb till after the complete formulation of all its members.

Fifth Hour.

Man stands upon his feet, he is weaned from earth, he walks and goes where he will.

Explanation.

The number five is that of the soul, typified by the quintessence which results from the equilibrium of the four elements; in the Tarot this number is represented by the high-priest or spiritual autocrat, type of the human will, that high-priestess who alone decides our eternal destinies.

Sixth Hour.

The animals pass before Adam, and he gives a suitable name to each.

Explanation.

Man by toil subdues the earth and overcomes the animals; by the manifestation of his liberty he produces his word or speech in the environment which obeys him; herein primordial creation is completed. God formed man on the sixth day, but at the sixth hour of the day man fulfils the work of God, and to some extent recreates himself, by enthroning himself as the king of nature, which he subjects by his speech.

Seventh Hour.

God gives Adam a companion brought forth out of the man's own substance.

Explanation.

When God had created man in his own image, He rested on the seventh day, for He had given unto Himself a fruitful bride who would unceasingly work for Him; nature is the bride of God, and God rests on her. Man, becoming creator in his turn by means of the word, gives himself a companion like unto himself, on whose love he may lean henceforth; woman is the work of man; by loving her, he makes her beautiful, and he also makes her a mother; woman is true human nature, daughter and mother of man, grand-daughter and grandmother of God.

Eighth Hour.

Adam and Eve enter the nuptial bed; they are two when they lie down, and when they rise they are four.

Explanation.

The tetrad joined to the tetrad represents form balancing form, creation issuing from creation, the eternal equipoise of life; seven being the number of God's rest, the unity which follows it signifies man, who toils and co-operates with nature in the work of creation.

Ninth Hour.

God imposes his law on man.

Explanation.

Nine is the number of initiation, because, being composed of three times three, it represents the divine idea and the absolute philosophy of numbers, for which reason Apollonius says that the mysteries of the number nine are not to be revealed.

Tenth Hour.

At the tenth hour Adam falls into sin.

Explanation.

According to the kabbalist ten is the number of matter, of which the special sign is zero; in the tree of the sephiroth ten represents Malchuth, or exterior and material substance; the sin of Adam is therefore materialism, and the fruit which he plucks from the tree represents flesh isolated from spirit, zero separated from unity, the schism of the number ten, giving on the one side a despoiled unity and on the other nothingness and death.

Eleventh Hour.

At the eleventh hour the sinner is condemned to labor, and to expiate his sin by suffering.

Explanation.

In the Tarot, eleven represents force, which is acquired through trials; God sends man pain as a means of salvation, and hence he must strive and endure that he may conquer intelligence and life.

Twelfth Hour.

Man and woman undergo their sentence; the expiation begins, and the liberator is promised.

Explanation.

Such is the completion of moral birth; man is fulfilled, for he is dedicated to the sacrifice which regenerates; the exile of Adam is like that of Œdipus; like Œdipus he becomes the father of two enemies, but the daughter of Œdipus is the pious and virginal Antigone, while Mary issues from the race of Adam.

FINIS.

Éliphas Lévi

A Word from the Publisher

Thank you for purchasing this book from The R.A.M.S. Library of Alchemy. During his lifetime, Hans Nintzel dedicated himself to the identification, acquisition, study, retyping and, when necessary, translation of what he considered to be the most important known works on Alchemy. Hans was assisted by his sparse network of fellow Alchemists, all members of the Restorers of Alchemical Manuscripts Society (R.A.M.S.). I was an active member of R.A.M.S.

Hans provided copies of the R.A.M.S. works as photocopies. My goal is to publish all of them as professionally printed books.

The works from the original R.A.M.S. Library are republished by R.A.M.S. Publishing Company in the collection, "The R.A.M.S. Library of Alchemy," with permission of the Estate of Hans W. Nintzel.

If you have a work on Alchemy that you believe should be a part of the R.A.M.S. Library, please contact me through R.A.M.S. Publishing Company.

Philip N. Wheeler

https://ramslibraryofalchemy.blogspot.com/

The R.A.M.S. Library of Alchemy

The study and practice of Alchemy was extremely important to Hans W. Nintzel. He assembled this Library over a period spanning more than three decades, guided by his teacher Frater Albertus. The R.A.M.S. Library of Alchemy includes all of the most valuable Alchemical texts that Hans painstakingly located, acquired, retyped, and translated during his lifetime, with help from other R.A.M.S. members.

The following is a list of the volumes that are currently available. Volumes that contain works from multiple authors may have only the principle author or editor listed.

Volume	Title	Author or Editor
1	Twelve Keys of Basilius Valentinus	Basilius Valentinus
2	Triumphal Chariot of Antimony	Basilius Valentinus
3	His Secret Book	Artephius
4	The Golden Work	Hermes Trismegistus
5	Three Works of Ripley	George Ripley
6	Four Works of Paracelsus	Paracelsus
7	Bacstrom's Notebooks, Part 1	Sigismund Bacstrom
8	Bacstrom's Notebooks, Part 2	Sigismund Bacstrom
9	Summa Perfectionis	Geber (Abu Musa Jabir ibn Hayyan)
10	The Five Centuries	Rudolph Glauber
11	The Greater and Lesser Edifyer	Johann Grashoff
12	Chemical Secrets and Experiments	Sir Kenelm Digby
13	The Turba Philosophorum	Arisleus
14	Das Aceton	Christian Becker
15	The Art of Distillation	John French
16	Non-Violent Destruction of the Atom	Nintzel & Wheeler
17	Philosophical Furnaces	Rudolph Glauber
18	The Last Will and Testament	Basilius Valentinus
19	TBD	
20	TBD	
21	Alchemical Symbols, Fourth Edition	Philip N. Wheeler

22	The Book of Formulas	John Hazelrigg
23	18 Short Tracts	Hans W. Nintzel
24	Bacstrom's Notebooks, Part 3	Sigismund Bacstrom
25	A Discourse on Fire and Salt	Blaise Vignere
26	The Mineral Work	Johan Hollandus
27	The Vegetable Work	Johan Hollandus
28	Lamspring's Process	Lamspring
29	The Book of Abraham the Jew	Abraham Eleazar
30	Five Short Works of Glauber	Johann Glauber
31	The Metamorphosis of the Planets	Johannes Monte-Snyder
32	Four Works of Roger Bacon	Roger Bacon
33	The Golden Chain of Homer	Homerus, Kirchweger, Nintzel, Wheeler
34	Alchemy Rediscovered and Restored	Archibald Cochren
35	Aurifontina Chymica	John Houpreght
36	The Golden Fleece	Salomon Trismosin
37	The Transmutation of Base Metals into Gold and Silver	David Beuther
38	Sanguis Naturae	Christopher Grummet
39	A Revelation of thye Secret Spirit	Giovanni Lambi
40	The Holy Guide, Part 1	John Heydon
41	The Holy Guide, Part 2	John Heydon
42	Secreta Alchymiae	Kalid Persica
43	The Golden Treatise of Hermes	Hermes Trismegistus
44	Potpourri of Alchemy, Part 1	Hans W. Nintzel
44	Potpourri of Alchemy, Part 2	Hans W. Nintzel
46	TBD	
47	Selected Chemical Universal and Particular Processes	Alexius von Ruesenstein
48	TBD	
49	Splendor Solis	Solomon Trismosin
50	Transcendent Magic	Eliphas Levi

http://ramsalchemy.jimdo.com

www.ingramcontent.com/pod-product-compliance
Lightning Source LLC
Chambersburg PA
CBHW062317220526
45469CB00008B/2540